Multivariate Analysis of Variance and Repeated Measures

A practical approach for behavioural scientists

Multivariate Analysis of Variance and Repeated Measures

A practical approach for behavioural scientists

D. J. HAND and **C. C. TAYLOR**

The Institute of Psychiatry
De Crespigny Park
London

London   New York
CHAPMAN AND HALL

First published in 1987 by
Chapman and Hall Ltd
11 New Fetter Lane, London EC4P 4EE
Published in the USA by
Chapman and Hall
29 West 35th Street, New York, NY 10001

Output from SPSS-X procedures is printed with permission of SPSS Inc.,
Chicago, Illinois 60611.

Output from SAS procedures is printed with permission of SAS Institute Inc.,
Cary, NC 27511-8000, Copyright © 1982.

Output from BMDP procedures is reproduced with permission of BMDP
Statistical Software, Los Angeles, California 90025.

Printed in Great Britain at the University Press, Cambridge

ISBN 0 412 25810 2   (hardback)
0 412 25800 5 (paperback)

British Library Cataloguing in Publication Data

Hand, D. J.
  Multivariate analysis of variance and repeated measures.
  1. Psychology—Statistical methods
  2. Analysis of variance
  I. Title   II. Taylor, C. C.
  519.5'352'0243   BF39

  ISBN 0-412-25810-2
  ISBN 0-412-25800-5 Pbk

Library of Congress Cataloging-in-Publication Data

Hand, D. J.
  Multivariate analysis of variance and repeated measures.
  Bibliography: p.
  Includes index.
  1. Multivariate analysis.   2. Social sciences—Statistical methods.
3. Analysis of variance.
I. Taylor, C. C., 1943–   . II. Title.
QA278.H345   1987   519.5'35   86-28421
ISBN 0-412-25810-2
ISBN 0-412-25800-5 (pbk.)

To Rachel and Emily

Contents

Contents

viii

Contents

Preface

Although there are many excellent books on multivariate statistics available, none of them is aimed at the mathematically naïve. This is not surprising since statistics is an inherently mathematical subject. To obtain a really effective grasp of statistics, and multivariate statistics in particular, one requires a good understanding of the mathematical languages needed to manipulate multidimensional concepts. A facility with matrix algebra, especially, is needed.

This poses a serious problem for the researcher. He or she may be an expert – perhaps even an internationally renowned expert – in some other domain of science but may not have the mathematical background to be able to acquire or understand the mathematics quickly and easily. From the point of view of such an expert the statistical concepts may be an extra burden to be learned – a burden because they are not his or her primary interest. And yet the concepts must be learnt if the researcher is to be able to formulate the multivariate questions or, at least, to be able to communicate in mutually intelligible terms with a statistician.

Worse is to come. Multivariate statistics is not practicable by hand. Its increasingly widespread use is very much a development of the computer age. Computers are essential for its use. This means that the researcher, having formulated a research question in the language of his or her own science, having learnt and reformulated it in the language of multivariate statistics, now has to learn to use one of the computer packages for tackling such analyses – not to mention the job control language needed to get the computer to accept the program.

The process has become a sequence of stages, each leading one step further from the researcher's primary interest, each requiring an investment of time and intellectual energy which the researcher may be unwilling to make.

This book aims to ease the difficulty at least for one particular multivariate technique, the use of which is growing dramatically. We mean, of course, multivariate analysis of variance.

When planning this book we had two distinct and to a certain extent conflicting aims in mind. The first was that we wanted to present a connected non-mathematical account of the concepts of multivariate analysis of variance which would be readily comprehensible to researchers in the behavioural sciences. The second was that we wanted to show such researchers how actually to do a multivariate analysis of variance. These aims conflict in that embedding numerical examples in the text can be disruptive to the flow of development of ideas and can sometimes be offputting to the sort of readers we had in mind. We therefore attempted to achieve the aims while avoiding these problems in two ways.

The first is the separation into the theoretical discussion of Part 1 and the applied examples of Part 2. The textual examples in Part 1 are all of an abstract form and the real examples of Part 2 show what happens with real data. Following the Contents is a table showing the types of analysis used in each of the studies in Part 2. The table is designed to act as a link to real examples of each sub-type of analysis. Thus one can either read the Part 2 examples in their entirety, as examples of complete analyses, or one can use the above-mentioned table to key into several different examples of a particular type of analysis.

The second way we have attempted to satisfy both aims is by inserting simple numerical examples into Part 1, set up as inset passages so as not to interrupt the theoretical development. They may thus either be read in parallel with the theory or skipped to be returned to later. The difference between these examples and those of Part 2, apart from one of simplicity, is that the Part 1 examples involve artificial data. We do not attempt complete analyses in these examples, simply using the data to illustrate isolated points. By the same token, we do not concern ourselves with the sorts of problems arising with real data which can interfere with what would in principle be an elegant analysis. These issues are the domain of Part 2.

The fact that the book is non-mathematical has its disadvantages. It means that readers will not be able to develop new multivariate techniques. They will not be able to examine critically any restrictions on the existing techniques in order to evolve new and more general ones. These sorts of things require a deep grasp of the background mathematics. What they will be able to do after reading the book – or, at least, such is our hope – is to understand the aims and objectives of multivariate analysis of variance, be able to formulate research questions in an appropriate form, be able to write computer programs using one of the existing multivariate packages (after studying the manual, of course) and, most important of all, be able to make sense of the output.

Multivariate statistics is built upon univariate statistics. A grasp of the former would be difficult to achieve without a grasp of the latter. The reader of this book is thus expected to have some familiarity with univariate statistics and with standard univariate analysis of variance and regression analysis in particular. We also expect the reader to know what a matrix is, but no more than this.

Perhaps we should also comment at this point that many readers will not find this book easy to read. Although it does not contain any mathematics, it should be read like a mathematics text. That means that it should be read, re-read and re-read again. It is the exceptional reader who will understand and retain everything in it at a single reading.

A second way in which the book is unusual is in its approach to the analytic ideas. Rather than beginning with overall F-tests (as most univariate treatments do) we begin with the constituent tests of the analysis and build up to overall tests. This is described in Chapter 2. We have found that non-mathematical students accept this approach much more readily than the alternative of beginning at the top with an overall test or with the general linear model.

We would like to thank the researchers whose data we have used for giving us their permission and their valuable time. We are very grateful to our colleagues, who read through the manuscript and gave us their constructive comments. Particular thanks are due to Dr Owen White of the Institute of Psychiatry and to Professor Henri Rouanet of the University of Paris, who provided many hours of fruitful discussion. We are also very grateful to Richard Stileman of Chapman and Hall, who forbore from beating us over our collective heads as deadline followed deadline, and to Jean Howard, Pat Davies and Linda Button for typing illegible scripts.

Of course, it is quite likely that in spite of our best endeavours mistakes remain in the text. If this is the case, each of the authors would like to make it clear that any passages still containing mistakes were written by the other author. The analyses in Part 2 of the text are not put forward as the sole solution to the questions the researchers were asking – other analyses are possible. We have included the data sets for each of the studies so that our readers can explore alternative approaches.

D. J. Hand
C. C. Taylor
Institute of Psychiatry
London University
1986

TYPES OF ANALYSIS USED IN EACH STUDY

Within each study (A–H), sub-analyses are indexed by a number.

Analysis type	Study							
	A	B	C	D	E	F	G	H
Univariate	1	2	1	.	.	.	.	.
Unstructured multivariate	2	.	2,3	1,2	.	1,2,3	.	.
Structured multivariate (1 factor)	3	1,2	4	.	.	.	1	.
Structured multivariate (2 factor)	.	.	.	.	1,2	.	.	1,2,3
Repeated measures	.	1,2	4	.	1,2	.	1	1,2,3
Single between-groups factor	1,2,3	1,2	.	1,2	.	.	1	.
Two between-groups factors	.	.	.	.	1,2	1,2,3	.	1,2,3
Three between-groups factors	.	.	1,2,3,4	.	.	.	.	.
Covariates	3	.	3	.	1	2,3	.	1

PART 1———————The theory

Introduction ————————————1

1.1 WHAT IT IS ABOUT

A good way to begin this introductory chapter is with an attempt at a definition: precisely what do we mean by multivariate analysis of variance? To answer this, first let us remind ourselves what ordinary (univariate) analysis of variance, often abbreviated to anova, does. Since this is supposed to be merely a reminder, if the reader has not encountered anova before, then a brief study of one of the many elementary statistics texts available will pay dividends. We recommend Hays (1963).

Analysis of variance is a statistical tool for studying differences between the means, on some particular variable, of distinct groups of subjects. The variable in question is often called the response or dependent variable. Presumably the reader will already have encountered the two-group t-test. Anova extends the two-group t-test to several groups. These groups might be the categories of a nominal variable (e.g. treatment type, socio-economic group, or religion) or they might have arisen as the result of a more complex classification (for example, they might be the result of a cross-classification of sex (two levels) by therapeutic regime (three levels, say), forming six groups in all). A researcher may be interested in knowing whether there are differences between particular groups, whether there are certain patterns of differences between groups, or perhaps whether there are any differences at all between groups.

From the above it will be clear that the subjects within each group represent just a sample of those that could have been chosen from that class (others could have been given each treatment, there are other subjects from the same socio-economic group that could have been chosen, etc.). We thus imagine a notional population for each group – and we distinguish between the sample and population means. We would like to make statements about differences between population means, but such statements necessarily have to be based on only sample evidence.

So much for anova; now what about *manova* – multivariate analysis of variance? Manova is anova in which the single response variable is replaced by several variables. These variables might be substantively different from each other (height, weight, IQ); or they might be the same substantive item measured at a number of different times (e.g. GSR (galvanic skin response) measured after six repetitions of a stimulus). This last is the essence of repeated measures problems, which are thus seen to be a special case of multivariate analysis of variance.

Presumably, whatever else is expected of manova, the reader expects it to be more complicated than anova. In some ways this expectation is fully justified, so a natural follow-up question to 'What is it?' would be 'Why should we use it?' or 'What is it used for?'

There are, roughly speaking, three answers to this question. Often a researcher will have several variables, each of interest in its own right, and will wish to explore each of them by itself. Each might be analysed at a conventional significance level (say 5%) but this can mean a very high chance of concluding that some relationship exists when in fact none does. Manova, and in particular the simultaneous test procedures described later, permit one to control this effect. This is the first answer to the question of when manova should be used. Like the other answers, we explore it in depth below.

The second answer is that manova should be used when interest lies, not in the raw variables as measured, but in some combination of them. A very simple example of this is a change score, this being a difference of two variables, a measurement at time 1 and a measurement at time 2.

Finally, the third answer is that manova should be used when interest lies in exploring between-groups patterns of differences on a set of variables *in toto*. The individual variables are of no intrinsic interest, it is their union which matters. In some cases, if differences are found then the global investigation may be followed up by specific ones which explore which of the variables matter, but this is of secondary interest. Global questions of this kind need an intrinsically multivariate approach. It can happen that no one of a set of variables shows any distinction between groups, whereas a suitable *combination* of variables distinguishes well. Identifying that suitable combination is a multivariate task.

Statistics as a formal discipline is driven by a number of motivating influences. Among these are the demands made by the areas to which it is applied. The behavioural sciences, especially, provide motivation for multivariate analysis of variance via questions of the types outlined above, in recognition of the fact that human beings are too complex to be described by single variables.

1.2 BEFORE WE START

The remainder of this chapter deals with some introductory details of which the potential user of multivariate analysis of variance should be aware. There is one vital point to be emphasized, however, which follows on from the preceding discussion. This is simply that even though multivariate analysis of variance is a sophisticated technique, it does not perform miracles. If the basic research questions are ill-conceived, if the variables being fed into the analysis are poorly chosen or extremely unreliable, then the analysis will be pointless. Just as good solid foundations are essential for a skyscraper, so sound basic variables and theoretical questions are necessary for effective use of multivariate analysis of variance.

At the very beginning, therefore, we would like to emphasize the importance of *looking at the data*. It is a recipe for disaster to transfer the numerical results, still wet from the experimenter's pen, directly into a complex multivariate analysis of variance computer program. This can all too often lead to absurd results which could have been detected and prevented with ease if the elementary precaution of examining the data beforehand had been followed. The grossest errors can be concealed behind the summary statistics of a multivariate analysis. (A typical example would be a measuring instrument malfunction in the middle of an experiment, which led to most of one group consistently scoring zero.)

Before going on we must say something about the terminology used in multivariate analysis of variance. Although any text must to some extent contain technical terms, we have tried to use a minimum. None the less, throughout this book we shall use a number of common words in a technical sense. For most of these it will be clear if and when the technical sense is intended. One that perhaps needs clarifying is the word *observation*. A multivariate observation on a particular subject is the entire vector or set of response variable scores measured on that subject. A single score is only a constituent part of the observation.

We are assuming a basic familiarity with significance tests. If the reader does not have this then perusal of an elementary text is essential. In particular, the reader is expected to understand the terms significance level, power, and error types I and II.

Another word that we shall use is *matrix*. A matrix is a rectangular array of numbers. For deep understanding of manova a thorough grounding in the manipulation of matrices is essential, but it is the essence of this text that we have tried to avoid this need. This means, of course, that there are aspects of manova that we cannot cover, but we believe that the concepts can be described without matrix algebra and, via the use of computer programs, effective and deep research analyses undertaken.

In particular, we shall occasionally refer to *covariance matrices*. Between each pair of variables in a study there will exist a covariance. Similarly, each variable has an associated variance. We can define a square matrix in which each row and each column corresponds to one of the variables and in which each entry is the covariance of its row variable and column variable. (The elements on the diagonal – for which the row variable and column variable are the same – are thus the variances.) This square matrix is called a covariance matrix.

The reader will recall that ordinary anova requires that the data satisfy certain basic assumptions. These are that the observations (which in the univariate case are each a single number) within each group should be derived from populations following normal distributions, that the variances within these populations should be identical, and that the observations are independently drawn, both between and within groups. Manova makes similar assumptions: independent observations sampled from multivariate normal populations with each group having the same population covariance matrix.

Introduction

In both univariate and multivariate cases some of these assumptions are more critical than others in terms of robustness of the techniques to departures from them. For example, with the exception of the requirement of independent sampling, they become less important for large sample sizes. If the groups have equal sample sizes then the analysis is relatively insensitive to departures from identical covariance matrices, but if they have different sizes then this assumption becomes more critical. The validity of assumptions can be tested prior to conducting the manova or anova analysis, of course, but if this is done one must be wary of rejecting what would have been a perfectly acceptable analysis as a result of an overly sensitive prior test.

Transformations can be used to make skew or otherwise non-normal distributions more nearly normal, though practically these are usually restricted to transforming the distribution of each variable separately and hoping that the joint distribution becomes approximately multivariate normal. Some authors rely on the robustness of the tests and, except in extreme circumstances, recommend no transformation. In such cases (and, in fact, generally) one should take observed significance levels as guidelines rather than hard-and-fast decision rules.

As well as the grouping of the subjects into different categories, one sometimes wants to explore or allow for the influence of properties of subjects which do not fall into natural groups; which are, in fact, continuous variables. This is done via analysis of *covariance* and multivariate analysis of covariance. This book also covers such techniques. A basic assumption is that the extra independent variables – the covariates – are error free. To the extent that this is not true, there will be a downward bias in the estimates of strength of relationship between the covariates and the response variables. With multiple covariates further problems can arise if they are highly correlated. These problems, and ways to tackle them, are described in good textbooks on regression.

Univariate analysis of covariance assumes that the regression coefficients of the response variable on the covariates are the same in each group. Multivariate analysis of covariance makes similar assumptions. In both cases tests of the reasonability of these assumptions can be made (see Chapter 4).

A point concerning not so much the assumptions of manova, but more the validity of the data, relates to extreme or outlying data points. Manova, in common with many other techniques, can be sensitive to outlying observations, so some authors have recommended formal methods of checking for and removing outliers prior to the analysis.

Some authors consider that the techniques described in this book should only be applied to interval scale data. Our belief, to the contrary, is that the validity of the statistical conclusions depends only on whether or not the numbers meet the distributional assumptions. However, having said that, we should note that interpretation problems may arise on data which are merely ordinal or to which an arbitrarily complex transformation has been applied.

1.3 FURTHER READING

This book is intended as an initial step in understanding multivariate analysis of variance. We can regard it as being followed by three successor steps, each leading to a more sophisticated grasp of the subject.

The second step consists of textbooks which begin to introduce the mathematical terminology and yet place great emphasis on interpretation and the problems of real data. These are often written by non-statisticians with a strong statistical background. Two good examples, both of which we recommend to follow the present work, are Harris (1975) and Tabachnick and Fidell (1983). These are general texts on multivariate analysis, but have good discussions of analysis of variance in particular.

The next step is to more-mathematical texts. These may well assume a background knowledge of mathematics to degree level and certainly require an advanced understanding of matrix algebra. Examples of general multivariate texts at this level which contain discussions of analysis of variance are Srivastava and Carter (1983) and Mardia, Kent and Bibby (1979). Finn (1974) and Bock (1975) are two other recommended books at this level which place greater emphasis on multivariate analysis of variance.

The fourth and final step is to research papers published in journals. Those in statistical journals will obviously require a strong mathematical background. Some other disciplines, however, notably psychology, also publish papers in this area and these tend to emphasize interpretation.

There is one other class of reading material which we must mention. To carry out a multivariate analysis of variance the reader will undoubtedly use a computer program, and must, therefore, read the appropriate manual. We analysed the examples described in Part 2 on a variety of packages and, since their number continues to grow, have refrained from describing any particular program. Suitable systems are the SPSS-X Batch System* (MANOVA) (SPSS-X Inc., 1983), the SAS® system (GLM) (SAS Institute Inc., 1982), BMDP Statistical Software (Dixon *et al.*, 1981), and MULTIVARIANCE (Finn, 1972).

Display 1 The effect of training and practice on creativity and problem solving

---

The inset numerical examples throughout Part 1 all refer to the same research design. We felt that this was a convenient way to proceed since it means that the reader only has to grasp a single basic design. So that we can illustrate different points we do two things: first, different examples will refer to different aspects of the design, and second, we show what happens if different numbers are used in the

---

*SPSS-X is a trademark of SPSS Inc. of Chicago, Illinois, for its proprietary computer software.
SAS is the registered trademark of SAS Institute Inc., Cary, NC, USA.

analysis. We believe the reader will find the result more straightforward to follow than if we had interrupted the flow by describing many different (synthetic) designs.

In this Display we describe the basic design. We remind the reader that these inset examples in Part 1 are artificial – and that when reading them it should be borne in mind that they are being used to illustrate isolated points, not entire analyses, as described in the Preface. This first inset example simply describes the study used to illustrate many points in Part 1.

A researcher is interested in exploring the effects of training and practice on creativity and problem-solving skills. He drew a number of subjects, some of each sex, and randomly allocated them to one of three groups. We shall refer to these as *condition* groups. They are defined as follows:

G1: the subjects in this group received lectures on lateral thinking and the theory of problem solving, but no guidance or assistance in actual problem solving.
G2: the subjects in this group were presented with a range of puzzles and problems, none requiring specialist knowledge, and spent their time attempting to solve them. They received no expert help.
G3: this was a control group. The subjects had not been exposed to the methods or objectives of the experiment beforehand.

Since some subjects within each of the three conditions were male and some were female, we could describe the study as having six groups in all.

The effect of the different conditions was gauged by measuring a number of different variables. There were four measures of problem-solving skills, these being averages of the time to complete a number of tasks under different conditions. The measures were as follows:

M1: was taken when there was no time pressure and when the problems were presented verbally.
M2: was taken when the subject was told that he or she would be penalized according to the time taken, and the problems were presented verbally.
M3: as M1, but the problems involved geometric shape manipulation.
M4: as M2, but the problems involved geometric shape manipulation.

We can thus think of the four problem-solving skill measures as forming a two-by-two classification of (time pressure/no time pressure) by (verbal presentation/geometric shapes).

In addition to the measures of problem-solving skills the experimenter took three different measures of creativity. We will label these C1, C2, and C3. They represent three different questionnaires in common use, which are purported to measure creativity.

Between-subjects———————2
analysis

2.1 INTRODUCTION

In this chapter we describe comparisons between distinct groups of subjects. All that we say applies equally to both univariate and multivariate cases, so we shall keep things as simple and familiar as possible by using only the language of the former. To translate the discussion into the multivariate case, simply replace words such as 'mean' and 'average' by phrases such as 'vector of means' and 'vector of averages'.

In view of this, an alternative title for this chapter might have been 'some aspects of univariate analysis of variance re-examined', where the word 're-examined' has been carefully chosen: we shall be looking at some univariate anova concepts which have probably already been encountered, but we shall be looking at them from an unusual perspective. We believe that our approach is more natural. (For the more advanced reader, we note that it avoids problems of using generalized inverses or of imposing constraints on overparameterized linear models.)

One standard way of describing analysis of variance is as a partitioning of the overall variation in the data into a component which is due to the differences between the groups and a remaining component which is due to the variation within the groups. This partitioning of variation or variance is what leads to the name *analysis of variance.* From this basic starting-point one can go on to look at various patterns of between-groups differences. Our approach begins at the other end. We *start* by exploring the particular patterns of between-groups differences that the researcher is interested in, and the notion of partitioning variation will not appear until Section 2.6.

In fact we shall describe our approach as starting yet one stage further back: we start by summarizing each group's scores into single values, and then we explore the particular patterns of differences between summary values in which the researcher is interested. We shall use the group's means as our summary measurements, so: we start by finding the best estimates we can of the group means and then we explore the particular patterns of differences that the researcher is interested in.

We begin, in the next section, with a very simple case, and then extend it later in the chapter to more-complicated situations. As it turns out, the simplest case contains most of the basic ideas.

2.2 ONE—FACTOR DESIGNS

A *factor* is a discrete (categorical) variable. In a between-subjects design the different groups correspond to the different levels or categories of the factor. A very simple example of a factor is sex, with just two levels. Individual subjects lie in one or other of these two categories. A more general example is hair colour, perhaps with levels black, brown, blond, red, grey and white. As a third example, consider the allocation of each subject to one of the drug dosage groups low, medium or high. Note that this factor is qualitatively different from the other two: it has a natural, inherent ordering. We shall see ways to take advantage of this property below. This factor also differs from the earlier ones in that its levels and thus the numbers in each group are defined by the experimenter. In the other cases the levels were imposed by nature.

In this section we study the case of groups defined by a single factor. More-complicated cases involving two or more factors, for example involving sex and hair colour classification simultaneously, will be dealt with later.

We shall spend the rest of this section introducing a way to describe the patterns of between-groups differences that we are interested in studying. Note that we are here only concerning ourselves with *describing* the patterns of differences. We are not (yet) addressing the questions of *estimating* the sizes of any such differences, or of *testing* the statistical significance of such differences. These two questions will be the subjects of later sections.

We start our process of description by introducing a technical term:

DEFINITION A *contrast* on a set of g groups is any set of g numbers which sum to zero.

The use of this idea in describing patterns of differences is best illustrated by way of a series of examples.

■ EXAMPLE 1 Suppose that we have two groups and we wish to compare them. That is, the 'pattern of differences' we wish to study is simply the difference between the two groups. As we have already remarked, we shall summarize the scores of each group by their means. Thus our aim here is to study the difference between the two group means. If m_1 is the mean of the first group and m_2 the mean of the second group, we wish to study $(m_1 - m_2)$. We can think of this as applying a weight $+1$ to the first group mean and a weight -1 to the second, and then examining the size of the *weighted sum of means*. The contrast describing the pattern of differences we are interested in is simply this set of weights. That is, the contrast is $(+1 \ -1)$ – this is a set of two numbers which sums to zero and so satisfies our definition of a contrast. The size of the difference in which we are interested is the *value of the contrast.* ■

This might seem a laborious way of describing what is simply the sort of situation tested by a two-sample t-test, but it has the merit of generalizing to more-complicated cases.

To summarize: in this two-group case we describe the pattern of differences that we are interested in studying by the contrast $(+1 \; -1)$.

■ EXAMPLE 2   In an experiment in which each of three treatment groups is to be compared with a fourth control group we might be interested in the following patterns of differences:

(a) The difference between the first treatment group and the control group.
(b) The difference between the second treatment group and the control group.
(c) The difference between the third treatment group and the control group.

The reader will recognize a certain similarity between each of (a), (b) and (c) and Example 1. In each of the former the pattern of differences to be studied is a simple difference between two groups. If we write the four groups in the order

       Treatment 1   Treatment 2   Treatment 3   Control

then we can use Example 1 to demonstrate that the following three contrasts describe the pattern of differences in (a), (b) and (c):

1(i)	1	0	0	−1
1(ii)	0	1	0	−1
1(iii)	0	0	1	−1

A zero, of course, indicating that the corresponding group mean has a weight of zero, shows that this group does not feature in this pattern of differences.

At this point we would like to emphasize again that this set of three contrasts (or, equivalently, the 3×4 matrix made up of the three contrasts) provides a *formal* or *symbolic description* of the pattern of differences between groups. It may or may not be appropriate to estimate the size of any such pattern by directly applying the weights to the group sample means and adding them together. This was acceptable in Example 1 but in general it is not. We shall discuss estimation below. Leaving estimation aside for the moment, the above matrix encapsulates the kinds of differences we wish to study. ■

To summarize: in general, patterns of differences in which we are interested can be represented by sets of contrasts.

■ EXAMPLE 3   In this example we continue with the four-group experimental set-up of Example 2. Further patterns of differences in which one might be interested in that design are:

(a) Do the three treatments have different effects? That is, are there any differences between the three treatment groups or are such differences zero?

11

Between-subjects analysis

(b) If (a) shows zero differences, so that the three treatments have the same effect, is this common effect different from zero? That is, is there any difference between the control effect and the common treatment effect?

The first of these questions can be approached in many ways. One way is summarized by the contrasts

2(i)	1	−1	0	0
2(ii)	0	1	−1	0

These describe, respectively, differences between groups 1 and 2 and groups 2 and 3. If the treatments have a common effect then these patterns of differences will both be zero. If, however, there are differences between treatments then at least one of the patterns of differences will be non-zero.

In the event that 2(i) and 2(ii) lead to zero values, question (b) makes good sense. The pattern of between-groups differences in this question can be written as

3(i)	1	1	1	−3

with the equal numbers signifying that these groups enter in an equivalent way into the particular difference pattern we are interested in.

Again we remark that this is merely a formal description of the pattern of differences. The best way to estimate the value of the difference represented by this pattern will be explored below. ∎

The reader might have noted some ambiguity in our definition of contrasts and the way we have described them above. This is simply that we could multiply each of the elements of a contrast by the same number and the result would be another contrast representing the same pattern of differences. Thus, in Example 1, we could use $(+1 \ -1)$ but we could equally use $(+5 \ -5)$ or $(36 \ -36)$. All of these satisfy our definition and express the required pattern of differences, but multiplied by a known figure. In Example 3 we could use the contrast 3(i) as we described above:

	(1	1	1	−3)
or we could use:	(1/3	1/3	1/3	−1)
or even:	(10	10	10	−30).

All of these are equivalent in terms of our definition above, the latter two giving values of the same contrast multiplied by 1/3 and 10 respectively. Note that when the pattern of differences gives a zero for the value of contrast 3(i) it also gives zero for the others.

Often it is convenient to rescale the contrasts so that their sum of squared elements equals one. Then the contrasts $(1 \ -)$ and $(1/3 \ 1/3 \ 1/3 \ -1)$ become instead $(1/\sqrt{2} \ -1/\sqrt{2})$ and $(1/\sqrt{12} \ 1/\sqrt{12} \ 1/\sqrt{12} \ -3/\sqrt{12})$ respectively. Such a rescaling process is called *normalization*.

2.3 COMPARISONS ON A SINGLE FACTOR

We begin this section with two definitions:

DEFINITION A *comparison* is the set of between-group differences which can be represented by a particular given set of contrasts.

DEFINITION A *complete comparison* for a factor is a comparison which represents all possible patterns of differences between its levels.

The reader may be puzzled by this second definition, wondering how such a comparison can be expressed, short of writing an infinite number of contrasts, one for each of the possible patterns of differences that could exist. Again, a series of examples seems the best way to proceed.

Beginning with the two-group case, we saw that there is only one pattern of differences possible: that associated with the contrast $(+1 \ -1)$ as described above. This means that for the two-group case this single contrast is also a complete comparison. In the three-group case, however, things become qualitatively different.

■ EXAMPLE 4   For a design with three groups, all of the contrasts

4(i)	1	0	-1
4(ii)	2	-1	-1
4(iii)	1	1	-2
4(iv)	3	-1	-2

are valid contrasts for describing patterns of differences. Also, and in particular, are

5(i)	1	-1	0
5(ii)	1	0	-1
5(iii)	0	1	-1

Furthermore, any of the contrasts 4(i) to 4(iv) can be obtained by suitably combining 5(i) to 5(iii), and rescaling if necessary. For example,

4(ii)	2	-1	-1

is given by adding

5(i)	1	-1	0

and

5(ii)	1	0	-1

where by 'adding' we mean that corresponding elements in each contrast are added. Similarly,

4(iii)   1   1   −2

is given by adding

5(ii)   1   0   −1

and

5(iii)   0   1   −1

A slightly more complex combination yields 4(iv). It is given by

$(5(i) \times 2) + 5(ii) + 5(iii),$

where the factor 2 represents a rescaling of contrast 5(i), that is, we replace $(1 \; -1 \; 0)$ by $(2 \; -2 \; 0)$ in the sum.

It is possible to show that *any* between-groups contrast (i.e. any pattern of differences between three groups) can be represented as a combination of 5(i), 5(ii) and 5(iii), rescaling if necessary.

This, then, is what we mean by a complete comparison. Used in conjunction 5(i), 5(ii) and 5(iii) can represent any pattern of differences between groups and so together they represent a complete comparison. ■

In this example we have listed all (three) pairwise contrasts – that is, all possible differences between pairs of groups. One might suppose from this that in general to describe a complete comparison one would need to list all such contrasts (with g groups, there would be $g(g-1)/2$ such contrasts). However, this is not the case. Indeed it is not even the case in the three-group example. If we look closely at 5(i), 5(ii) and 5(iii) we see that

$5(iii) = 5(ii) - 5(i)$

Thus the third contrast is superfluous. In fact, since we could equally have written 5(i) or 5(ii) in terms of the others we see that any one of the three can be regarded as superfluous. The point is that we do not need all three – any two will do.

Extrapolating this argument to g groups, we have that $(g-1)$ suitably chosen different pairwise contrasts form a complete comparison. Any pattern of differences can be explained in terms of these contrasts. These contrasts summarize all the information there is about between-groups differences.

Note the phrase 'suitably chosen' in the above. Any arbitrary set will not do. For example, it is clear that the set

6(i)   1   −1   0   0
6(ii)   1   0   −1   0
6(iii)   0   1   −1   0

does not permit us to express every pattern of differences on four groups. There is no information in this set about the fourth group and no combination of 6(i) to 6(iii) will tell us anything about differences involving the fourth group. (This contrast set 6 represents an *incomplete comparison*.)

A more subtle example is

7(i)	3	−1	−1	−1
7(ii)	−1	2	−1	0
7(iii)	1	3	−3	−1

The reader might spend some time (but only a little) trying to use this set to yield the difference between the third and fourth groups.

Two questions thus arise in connection with complete comparisons. Firstly, how do we choose a suitable set – how can we choose contrasts so that they form a complete comparison? And secondly, must this set involve only pairwise difference contrasts?

Taking the second question first, the answer is no. For example, the set

8(i)	1	−1	0	0
8(ii)	1	1	−2	0
8(iii)	1	1	1	−3

permits us to express any between-groups difference on the four groups. Thus we can obtain the difference between groups 2 and 3 as

$$(0 \quad 2 \quad -2 \quad 0) = 8(\text{ii}) - 8(\text{i}).$$

We recall here the arbitrariness of rescaling – so that

$$(0 \quad 1 \quad -1 \quad 0) \quad \text{and} \quad (0 \quad 2 \quad -2 \quad 0)$$

are equivalent.

There is clearly a large number of possible sets of contrasts each of which permits us to express any possible pattern of between-groups differences. That is, a complete comparison can be written in many ways, using many different contrasts (except in the two-group case). In fact, it can be done in an infinite number of ways.

However, each of these sets need contain only $(g-1)$ contrasts (one less than the number of groups, g). We can physically write a complete comparison using more than $(g-1)$ contrasts but if we do so, then some are superfluous and we can reduce the number to $(g-1)$ without losing the ability to represent differences.

Turning now to the first question above, how do we choose a suitable set, and how do we identify whether or not a set is a complete comparison? The answer is more complicated. It is that a set of $(g-1)$ contrasts forms a complete comparison on g groups if each of them provides information about a pattern of between-groups differences which cannot be obtained from the others. Thus in the set

9(i)	1	-1	0	0
9(ii)	1	0	-1	0
9(iii)	1	0	0	-1

if we drop any one contrast we are completely losing information about one group. Then, certainly, differences involving that group cannot be expressed in terms of the remaining contrasts.

A more interesting example arises from contrast set 8 if we drop the first contrast, leaving

| 8(ii) | 1 | 1 | -2 | 0 |
| 8(iii) | 1 | 1 | 1 | -3 |

Now no information remains about differences between group 1 and group 2.

Thus the answer to the question 'how do we identify whether or not a set of contrasts forms a complete comparison?' is: firstly, if there are g groups there must be at least $(g-1)$ contrasts; and secondly, none must be expressible in terms of the others. (The technical term for such a situation is that they must be *linearly independent*.)

Of course, whether or not the contrasts form a complete comparison, the answer to the general question 'how do we choose a suitable set of contrasts?' is that it depends on the research questions – it depends what patterns of differences the researcher wishes to investigate – and we shall see in the following sections that it may or may not require a complete comparison.

We can deduce from all the above that an incomplete comparison is a complete comparison from which some necessary contrasts have been omitted. Thus whereas a complete comparison permits any between-groups pattern of differences to be described, an incomplete comparison does not permit description of certain types of patterns of differences. We can put this conversely by saying that an incomplete comparison only permits certain types of patterns to be described.

Finally, to conclude this section, another definition:

DEFINITION The number of *degrees of freedom* in a comparison is the minimum number of contrasts needed to represent that comparison.

In particular we see that a complete comparison on g groups has $(g-1)$ degrees of freedom. The reader will presumably have already encountered the idea of degrees of freedom as a quantity pertaining to the variance within a single group or sample (as for instance in a t-test). The above definition merely extends the idea from variation within groups (the degrees of freedom within groups) to differences between groups (the degrees of freedom for a comparison). We will use these ideas again in Section 2.6.

2.4 ESTIMATING THE SIZES OF PATTERNS OF DIFFERENCES

In Sections 2.2 and 2.3 we went to great pains to emphasize that we were only talking about ways to *describe* patterns of differences. We outlined the descriptive ideas in some depth because of their central importance. It is only when one has a precise description, identifying exactly what one wishes to study, that it is legitimate to go on to estimation of size, and then to testing to see how that size compares with some hypothesized value. (This might be a general scientific dictum: first one must say exactly what the phenomenon being studied is; only then can one measure it; and only then can one compare its size with other sizes.) In this section we shall be concerned with how to estimate the size of the population differences underlying the observed sample means.

A complete comparison represents all possible patterns of differences between groups. That is, as we have discussed above, the contrasts describing the comparison can be combined so that *any* particular pattern of differences can be represented. An incomplete comparison is incomplete precisely because there are certain patterns it does not represent and which it cannot represent, no matter how its describing contrasts are combined. We can phrase this in a complementary way by saying that if we describe the relationships between the means of a set of groups using an incomplete comparison we are assuming that certain kinds of differences are zero. An example will clarify this.

■ EXAMPLE 5   The set

10(i)	1	-1	0
10(ii)	1	1	-2

is a complete comparison for three groups. If we use the sample data to estimate the sizes of the differences represented by these contrasts then we are not assuming anything a priori about the relationships between the true population means underlying the groups. However, if we use the incomplete comparison consisting of the single contrast

11(i)	1	1	-2

we are not allowing for any difference between groups 1 and 2. That is, we are assuming groups 1 and 2 have the same true population mean. Their sample means may differ, of course, but by not including $(1\ -1\ 0)$ in the comparison we are saying that we believe the difference between the population means is zero (and the sample mean difference, if any, is due to random fluctuation). That is, we are expressing a constraint or restriction on what we believe about the true population means. ■

When we use a complete comparison, since we are assuming no restrictions on the patterns of relationships between groups, each group's sample mean provides the best estimate of that group's population mean. Thus, given that we have a complete comparison, the between-groups differences can be estimated by applying the numbers in the contrasts directly to the group sample means.

An incomplete comparison, however, imposes a restriction, and this restriction implies that we might have more information available about a group's true population mean than just its own sample mean. In the above three-group example, both group 1 and group 2 can be used to estimate the population mean of group 1 (or group 2) since these groups are assumed to have the same population mean.

Moreover, if groups 1 and 2 had different sample sizes we would clearly want to assign more credibility to the larger group. To take an extreme, suppose group 1 had 10,000 sample observations and group 2 had only 1. The former would yield a much more precise estimate of the (assumed common) mean, so that when estimating this common mean it would be unreasonable to give the two sample means equal weights. It would be more appropriate to weight group 1 10,000 times as heavily as group 2.

In general, then, when using an incomplete comparison we are implicitly assuming that some patterns of differences are zero and in this case the best estimates of the underlying true population means are not (except in some special, simple cases – see the next section) the corresponding sample means, but are some more complicated function of all the sample means. Fortunately these more complicated expressions are worked out automatically by the computer program from a specification of the contrasts and the group sizes.

We have seen that the way the value of a contrast is estimated depends on whether or not the comparison of which it is a part is complete or in some way incomplete. We can generalize this. For two incomplete comparisons where one includes the second (i.e. not all of the patterns of differences in the one can be derived from the second) any common contrasts will usually have different estimated values. An explanation of the qualifying 'usually' occurs in the next section.

Since incomplete comparisons correspond to assumptions that certain paterns of differences are zero, when might it be appropriate to use incomplete comparisons? There are two basic situations. First, when one has strong a priori beliefs that some patterns of between-groups differences are zero – perhaps there is a large mass of evidence from other studies. And, second, when one has tested a pattern and found that it is not significantly different from zero. We shall examine the testing procedure in more detail in a later section.

The second of these two situations is the more important and interesting case. If often occurs when one has a set of contrasts which lie in some kind of natural order or hierarchy and for which those early in the order only make sense if higher-order ones are zero. Presented with such a case a natural procedure is to begin with the complete comparison and examine the highest-order contrast. If this is found to be zero one moves on to examine the incomplete comparison with this highest-order contrast omitted. This process is continued in a sequential manner until a non-zero contrast is found.

The alternative to an ordered or hierarchical structure of contrasts is the case of an unordered set. For example, when the researcher is simply interested in a set of different unrelated hypotheses (contrast set 9 provides an example). In this case it is appropriate to examine each contrast as part of a complete comparison. We are then simply exploring the size of each pattern of differences without making assumptions about other patterns. But if one should be found to be zero there is nothing to stop us working with an incomplete comparison for the other contrasts. Omitting a zero value contrast could lead to more-efficient estimates of the others, as our example above shows.

We can summarize this by saying that if certain patterns of differences are zero (i.e. if certain contrasts have zero value) then one can get more-efficient estimators than through direct application of contrast weights to sample means. However, these more efficient estimators are based on the assumptions that the specified contrasts have zero values. If this assumption is wrong then poor estimates, even misleading and meaningless ones, could result. In particular, a common and incorrect procedure is to assume that high-order interactions (p. 33) are zero simply to reduce the size of the problem. Often it would be far better to use the estimates based on the complete comparison – the straightforward differences between the sample means. We shall return to this topic in Section 2.12.

Display 2 A single-factor, three-level univariate experiment

This example uses only variable M1 from Display 1. The experimenter wished to compare the three condition levels using only this problem-solving measure: in particular, to compare conditions 1 and 3 (the lectures versus the control) and conditions 2 and 3 (the practice versus the control) to see if either of the 'treatment' conditions produced any effect. The data collected are shown in Table X.2.1. You will notice that there are different numbers of subjects in each group. Note also that in this simple preliminary analysis no account has been taken of sex – the researcher is ignoring any possible sex effect. We shall see the consequences of this in a later Display.

Writing the conditions in order G1, G2, G3, it is clear that contrasts describing the patterns of differences the researcher is concerned with are

| 1 | 0 | -1 |
| 0 | 1 | -1 |

If we use these two contrasts simultaneously, we have a complete comparison. In this case the estimated population means are just given by the sample means: 11.0, 13.0, 7.0 respectively. These yield estimated contrast values of

$$(11 \times 1) + (13 \times 0) + (7 \times (-1)) = 4$$

and

$$(11 \times 0) + (13 \times 1) + (7 \times (-1)) = 6$$

19

for the two contrasts above. Suppose, however, we estimate these contrasts separately, as incomplete comparisons. Taking the first contrast first, this means that we are regarding the second as if it had value zero – as if we know that G2 and G3 had identical means. (Our data in fact suggest this assumption is erroneous.) In this case we estimate the mean of G1 to be 11.0, as before, and the mean of G3 to be 9.5, it now being based on all the G2 and G3 observations. The estimated value of

Table X.2.1 The data for Display 2.

G1:	8, 7, 12, 11, 15, 13
G2:	10, 10, 10, 11, 9, 13, 18, 18, 12, 19
G3:	6, 6, 6, 10, 6, 9, 4, 5, 6, 7, 7, 11, 12, 3

the first contrast is then $(11 \times 1) + (9.5 \times (-1)) = 1.5$. Similarly, the value of the second contrast, when regarded as composing the entirety of an incomplete comparison, is 4.8. These estimated values are different from the ones using the complete comparison. Although in this simple example the calculations can be done by hand, in general the values of means and contrasts in incomplete comparisons are difficult to calculate and the work is best done by computer.

Display 3 Another pair of contrasts

Referring to the data of Display 2, a different experimenter is primarily interested in comparing groups G1 (lectures) and G2 (practice). Should these prove to have the same effect then the question of interest is whether their common score differs from that of the control. Contrasts for this case are

1	−1	0
1	1	−2

Taking the pair as a complete comparison necessarily gives the same estimated mean values as before, namely 11.0, 13.0 and 7.0 respectively for the three groups. From these the contrast values are estimated to be

$$(11 \times 1) + (13 \times (-1)) + (7 \times 0) = -2$$

and $\quad (11 \times 1) + (13 \times 1) + (7 \times (-2)) = 10$ respectively.

The complete comparison, however, is not what was wanted. The researcher first wanted to know the size of contrast $(1\ -1\ 0)$, making no restrictive assumptions. This is indeed the estimated contrast value obtained from the complete comparison. And then, if this was zero (in fact we estimate it to be −2) the researcher would explore the size of the $(1\ 1\ -2)$ contrast taking the first contrast to be zero. That is, taking the first two groups as having a common mean. This gives $(12.25 \times 1) + (12.25 \times 1) + (7.0 \times (-2)) = 10.5$ (the 12.25 is $[(11 \times 6) + (13 \times 10)]$, $(6 + 10)$).

2.5 A SPECIAL CASE

At several points in the preceding discussion we hinted that there were some situations in which things simplified. In these situations the best estimate of a pattern of differences is given simply by applying the contrast weights to the sample means, even though other patterns are assumed to have a zero value. Clearly it would be useful to know what these situations are, if only because then one can easily calculate sizes of between-groups differences by hand, avoiding the use of a computer. To describe these situations we need a further, technical definition:

DEFINITION Two contrasts are *orthogonal* to each other if their respective weights

$(a_1, a_2, \ldots, a_g)$ and $(b_1, b_2, \ldots, b_g)$

satisfy the condition that

$a_1 b_1/n_1 + a_2 b_2/n_2 + \ldots + a_g b_g/n_g$

sums to zero, where $n_1, n_2, \ldots$ are the sample sizes for the groups.

We shall give an example to illustrate this definition.

■ EXAMPLE 6   Note that if all the groups have the same sample size, then this simplifies to the condition

$a_1 b_1 + a_2 b_2 + \ldots + a_g b_g = 0$

For example, if there are three groups and they have equal-sized samples, then the following two contrasts are orthogonal to each other:

12(i)	1	0	−1
12(ii)	1	−2	1

since

$(1 \times 1) + (0 \times (-2)) + ((-1) \times 1) = 0.$ ■

Moving now to the case of four groups and equal sample sizes, the contrasts

13(i)	−2	−1	1	2
13(ii)	−1	1	1	−1

are orthogonal to each other. It is in fact always possible to find $(g-1)$ mutually orthogonal contrasts in g groups. (Thus sets of orthogonal contrasts are a special case of sets of linearly independent contrasts.) The reader might like to try to find a third contrast, orthogonal to 13(i) and 13(ii).

Now we come to the special situations referred to above: the best estimator of a difference pattern given by one contrast is unaffected by whether or not we assume the difference pattern given by a second contrast to be zero *provided these two contrasts are orthogonal*.

In terms of comparisons: if an incomplete comparison is formed by dropping a contrast which is orthogonal to those remaining, then the best estimates of the values of those remaining contrasts are unaltered.

Suppose, for example, we have the comparison

14(i)	1	-1	1	-1
14(ii)	1	1	-1	-1
14(iii)	1	-1	-1	1

and for simplicity assume that the design consists of equal-sized groups. Then these contrasts are all mutually orthogonal (the reader should verify this). We already know that, since 14(i) to 14(iii) form a complete comparison, the best estimators of these patterns of differences are given by applying the contrast weights directly to the group means. Now we are saying that, by virtue of the orthogonality of 14(iii) to the remaining two contrasts, if it is dropped the best estimators of 14(i) and 14(ii) are still obtained in this way, giving the same values as before. This is in distinction to the earlier situation, where dropping a contrast imposed constraints which meant that more-effective estimators could be obtained. We can go further. If we now also drop 14(ii), since 14(i) and 14(ii) are orthogonal the best estimator of 14(i) remains as before.

Had the groups not had equal sample sizes then, as the definition of orthogonality shows, contrasts 14(i) and 14(ii) might well not have been orthogonal. They are orthogonal for some patterns of sample sizes and not for others – it would be enlightening for the reader to carry out some arithmetic experimentation.

The reader who has already been exposed to univariate analysis of variance will probably have met the simple case of equal cell sizes and orthogonal contrasts. Many elementary texts introduce the subject this way for the good reasons that (a) the computation is more tractable and can be done by hand and (b) the theory simplifies (more on this when we discuss testing of patterns of differences). However, we strongly believe that that way of introducing the subject is not a good idea. Firstly, it is not necessary – computer programs can nowadays do the arithmetic, so that hand calculation is irrelevant. Secondly, and more importantly, it constrains the researcher to pose questions to fit the mould of his research tools, perhaps even restricting the questions he can ask to certain types (that is, difference patterns corresponding to orthogonal contrasts). This is not what statistics should do. If a researcher has clear and well-defined research questions, then statistical methodology should provide tools to answer those questions and not instead answer related, but different, questions. Of course, if one's questions naturally fall into a pattern of orthogonal contrasts then one may well wish to take advantage of the simplifications which result.

Display 4 Equal numbers of subjects and non-orthogonal contrasts

---

Another experimenter conducts an identical experiment to that of Display 2, but this one arranges things so that there are equal numbers of subjects under each condition. The data are given in Table X.4.1. This researcher is again interested in the contrasts

$$
\begin{array}{ccc}
1 & 0 & -1 \\
0 & 1 & -1
\end{array}
$$

Table X.4.1 Data for Display 4.
Note that unlike Table X.2.1, this set has equal numbers of subjects under each condition.

G1: 8, 7, 12, 8, 10, 11, 15, 13, 13, 13
G2: 10, 10, 10, 11, 9, 13, 18, 18, 12, 19
G3: 6, 6, 6, 10, 7, 9, 9, 4, 5, 8

Regarding these as a complete comparison (taking them together) gives estimated means of 11, 13 and 7 respectively, these being the simple cell means. (We contrived the data so that the equal and unequal sample size cases had equal cell means.) These again yield contrast values 4 and 6.

Now regarding each contrast as an incomplete comparison in its own right, just as in Display 2, we get for contrast $(1\ 0\ -1)$ an estimated value of 1 and for contrast $(0\ 1\ -1)$ an estimated value of 4. These are very different from those based on the complete comparison, even though there were equal numbers in each group.

Display 5 Orthogonal contrasts

Referring to the data of Display 4, another researcher wishes to examine contrasts

$$
\begin{array}{ccc}
1 & -1 & 0 \\
1 & 1 & -2
\end{array}
$$

This might be the researcher of Display 3, but he now has Display 4's data to work with – a data set which has equal numbers of subjects in each group. These contrasts are thus orthogonal.

If a complete comparison is used we will get the same mean estimates as in Display 4, leading to contrast values

$$
(11 \times 1) + (13 \times (-1)) + (7 \times 0) = -2 \quad \text{and}
$$
$$
(11 \times 1) + (13 \times 1) + (7 \times (-2)) = 10.
$$

But what about incomplete comparisons? The researcher of Display 3 will be interested in the incomplete comparison using the second contrast alone. This incomplete comparison assumes groups 1 and 2 have the same mean, namely $[(11 \times 10) + (13 \times 10)]/(10 + 10) = 12$. The second contrast then has value $(12 \times 1) + (12 \times 1) + (7 \times (-2)) = 10$. This is the *same* as the complete comparison result.

This illustrates that a contrast's estimated value is unaffected by whether or not contrasts orthogonal to it are constrained to be zero.

2.6 TESTING SINGLE CONTRASTS

Suppose that we are estimating a certain pattern of differences by applying the elements of a contrast directly to the sample means. This would be legitimate, for example, if the contrast was part of a complete comparison. Focusing attention on this estimated value for this particular pattern of differences, we now wish to conduct a statistical test. That is, we wish to compare the estimated value with an hypothesized value to see how unlikely it is for the observed discrepancy to occur solely by chance. Usually the hypothesized value is zero, though this need not always be the case.

The test statistic for such a situation is derived in exactly the same way as the test statistic in a single-sample t-test. In the latter we take the difference between the sample mean and the hypothesized mean and compare this difference with an estimate of its standard error. In the former we take the difference between the sample estimate of the pattern of differences and its hypothesized value and compare this difference with an estimate of its standard error. In each case we relate the calculated ratio to tables of the t-distribution.

Alternatively, and more usefully for our purposes, it will be recalled that the square of a t-statistic is an F-statistic. Thus, instead of comparing the difference between the sample and hypothesized values with the standard error, we could compare the square of this difference with the variance, and then relate this ratio to F-distribution tables.

The reader will recall that in Section 2.2 we noted that our definition of contrast was imprecise in that the same pattern of differences was represented by any arbitrary rescaling of the contrast's terms. We shall now make use of this fact, introducing a particular scaling so that we can describe the test statistic in a way which permits ready generalization.

First note that the variance of the estimate of a contrast is a weighted sum of the variances of the estimates of each group mean. Since, moreover, a standard assumption of analysis of variance is that each group has the *same* variance, this reduces to the fact that *the variance of the estimated contrast is proportional to the within-group variance.* (The constant of proportionality can be calculated from the contrast weights for each group, a_i, and the group sizes, n_i, as the sum over all groups of the terms a_i^2/n_i.)

Now we exercise our freedom of choice in rescaling the contrast. We choose the rescaling so that the constant of proportionality becomes 1. This means that the variance of the estimated contrast (using this rescaling) is equal to the within-groups variance. Our test statistic then simply becomes a ratio of the squared value of the rescaled contrast to the within-groups variance. The merit of all this rather complicated rescaling in a rather simple F-test is that it allows the following generalization.

The statistic is sometimes described as involving a ratio of a term measuring the between-groups pattern of differences to a term measuring the within-group variation. In the next section we shall generalize this to more than one contrast, so that the numerator of the ratio, instead of being a single squared term, becomes a sum of squared terms. The traditional way of describing this more general situation is that our test statistic then involves the ratio of the *hypothesis sum of squares* to the *error sum of squares*. Alternatively, when these are adjusted by dividing by their respective degrees of freedom to take account of the number of terms involved, the test statistic is the *hypothesis mean square* in ratio to the *error mean square*. Other names are *attributable variation* and *unexplained variation* respectively.

Display 6 Testing a single contrast

Referring to the example in Display 2, we have already estimated the sizes of the two contrasts. Now suppose we wish to test them. That is, we wish to see how likely it is that sample contrasts of such sizes could arise by chance if they had true population values of zero.

Taking the contrast $(1\ 0\ -1)$ to illustrate, as explained in the text, the test involves calculating the ratio of the squared contrast value to the within-groups variance – when the contrast has been suitably rescaled. Using the complete comparison result, this contrast had value 4. The rescaling involves dividing the original squared contrast value $(4^2 = 16)$ by the sum of the a_i^2/n_i values. In our case, using the Display 2 data, this sum is $(1/6 + 0 + 1/14) = .238$, so that the squared rescaled contrast is $16/0.238 = 67.2$.

Since the comparison involves three groups, and since one of the assumptions of anova is that the groups have identical population variances, the best estimate of within-group variance is obtained by calculating the average variance of the three groups. The computer does this, taking into account the sizes of the groups. Note that although each contrast only involves two groups, all three are used when calculating the best possible estimate of within-group variance. (If one ignored the middle group, which has a weight of zero in the first contrast, one would be conducting a two-sample t-test, for which the variance estimate is based on only the first and third groups. Thus a t-test could yield a different result. We are getting a better estimate of within-group variance, but at the cost of the assumption that the three groups have identical variance, when we estimate the variance as part of a

Table X.6.1 Anova output for tests on the two contrasts given in Display 2, as part of a complete comparison.

	Sum of squares	d.f.	Mean square	F	P
Constant	2842.394	1	2842.394	286.361	.000
Contrast 1	67.200	1	67.200	6.770	0.15
Contrast 2	210.000	1	210.000	21.157	.000
Within groups	268.000	27	9.926		

three-group anova.) The estimate of the common within-group variance turns out to be 9.926.

Thus, finally, the value of the test statistic is $(67.2/9.926) = 6.770$. This is compared to an F-distribution with 1 and 27 degrees of freedom. The first of these numbers is the number of independent contrasts we are testing (in this case we are examining a single contrast) and the second is the number of independent elements used in estimating the common variance. This is $(6 - 1) + (10 - 1) + (14 - 1) = 27$. Further details of this are given in texts on univariate anova. Table X.6.1 shows a typical form of output from a computer program testing each contrast separately as part of a complete comparison. Our result above is significant at the 5% level, but not at the 1% level.

So far in this section we have dealt with a special case: the case when the pattern of differences may be estimated by applying the contrast weights directly to the cell means. In the next section we turn to more general cases: cases of incomplete comparisons so that the raw contrast weights for a particular pattern of differences must generally be modified (by a computer program) before applying them to the cell means, as discussed in Section 2.4. Under these circumstances the same reasoning applies unchanged for testing single contrasts. However, before moving on to discuss the testing of comparisons formed from sets of contrasts, a few further remarks are in order about the error sums of squares used in these tests.

As we described at some length in Section 2.4, use of an incomplete comparison implies an assumption that certain patterns of differences are zero. These patterns may, of course, exhibit some random fluctuation about their true values of zero, but normally this will be small. However, and here is the important point, this *is* random fluctuation and so could legitimately be added in some way to the error sum of squares in the denominator of the F-test. Many computer programs permit this as an option. That is, they let one use either the straightforward *within-group* variation or the *pooled error* variation: that is, the variation due to within-group variation plus the observed random departure from zero of those patterns assumed to be zero. Which is better depends on the circumstances. Regarding departures from zero as due to random error might be thought risky – but one would not be doing this unless one had a strong belief that the underlying population difference patterns were zero. On the other hand, it is true that no pattern of differences is *exactly* zero. Pooling also tends to reduce the estimate of error variance and so to increase power, the chance of detecting real effects. One rationale for a choice would be based on the size of the estimated effect in relation to the size of the within-group variance.

■ EXAMPLE 7   At this point an example using a common one-factor situation is appropriate. We shall consider a case where the contrasts fall into a natural order.

Consider a study of the attitudes of people born in different decades. Suppose we have four groups, those born in the 1940s, the 1950s, the 1960s and the 1970s, and we wish to know how a certain attitude changes with birth decade. We might use the following contrasts to address this question:

15(i)	-3	-1	1	3
15(ii)	-1	1	1	-1
15(iii)	-1	3	-3	1

These represent, respectively, linear, quadratic and cubic patterns of change over time. That is, 15(i) represents a steady change over time at a constant rate, 15(ii) represents a time change in which the rate of change itself increases or decreases at a constant rate, and 15(iii) represents a more complex pattern of change with decade in which this latter rate 15(ii) itself changes (more on this topic later).

Since 15(iii) corresponds to the most complex pattern of differences we might first use the complete comparison defined by 15(i) to 15(iii) to test 15(iii) to see if it is zero. This is an example of an ordered situation, as mentioned in Section 2.4. If 15(iii) seems unnecessary then we might drop it (i.e. assume the pattern of population group mean differences represented by 15(iii) is really zero) and concentrate on the incomplete comparison given by 15(i) and 15(ii). This process can then be repeated, examining contrast 15(ii). At each step one can, if one wishes, pool the variation in those contrasts that one is assuming to be zero with the within-group variance, as described above. ■

2.7 TESTING COMPARISONS

As we have remarked, the reader will probably have already encountered the F-test in univariate analysis of variance as a test comparing the overall effects of a factor, rather than the case described above of testing single contrasts. We shall now extend our single contrast case to include this situation.

The attributable sum of squares used in testing a single contrast is a measure of the strength or size of that particular type of difference between the groups. When we come to testing the strength or size of a comparison – that is, of a set of contrasts – we need to calculate some overall attributable sum of squares for that comparison. That is, a sum of squares which can be attributed to the patterns of between-groups differences implicit in the whole comparison. Unfortunately it is seldom the case that, for the contrasts we wish to examine, we can derive this total attributable sum of squares by simply adding together the sums of squares due to each component contrast. This is only possible in certain special cases, which we shall describe below. Before we do so, however, the reader might care to glance back at Section 2.5 and speculate on what these special cases might be.

In the general case, when the attributable sum of squares is not simply the sum

Between-subjects analysis

of the separate contrast squares, the calculations involved in finding the overall attributable sum of squares can be quite complicated. Since this book is based on the assumption that the reader has available a computer program which will do the calculations automatically, we are going to make no attempt to describe how to do it. Readers who wish to pursue the matter may do so via the references in Sections 2.16 or 1.3.

The special case referred to above, where straightforward addition suffices, is, as the reader may have suspected, the case of orthogonal contrasts. For a comparison, complete or incomplete, which is expressed as a number of orthogonal contrasts, the overall attributable sum of squares due to the comparison is equal to the sum of the separate squares due to each contrast. In such a case one could calculate the individual squares by hand and add them to give the total, but any computer program will do it just as effectively for this case as for the non-orthogonal case.

Display 7 Testing a comparison

In Display 6 we tested each of the contrasts separately. Now suppose we are interested in testing the comparison formed as a combination of the two contrasts. In this case, with only three groups, it means that we are talking about a complete comparison, but the result applies to comparisons generally.

The overall sum of squares for this comparison is 220.800. We see by comparing this with Table X.6.1 of Display 6 that this is not the result obtained by adding the sums of squares for the two contrasts. The within-groups sum of squares remains the same as in Display 6 for the reasons explained there. Table X.7.1 shows the analysis of variance table for this test on the complete comparison.

As another example, consider the data and contrasts of Display 5. This had equal sample sizes in each of the three groups and it had orthogonal contrasts. If testing each contrast separately, but as part of a complete comparison, the anova table is as shown in Table X.7.2, and if grouping the two contrasts together and just testing the comparison we get Table X.7.3. (The test here is of whether there is any condition effect, without exploring details of precisely what it is.) Comparing Tables X.7.2 and X.7.3 we see that the individual contrast sum of squares does add to the total sum of squares. It is the orthogonality which causes this result.

Table X.7.1 Anova for complete comparison using unequal cells and non orthogonal contrasts.

	Sum of squares	d.f.	Mean square	F	P
Constant	2842.394	1	2842.394	286.361	.000
Comparison	220.800	2	110.400	11.122	.000
Within	268.000	27	9.926		

Table X.7.2 Anova table for equal cell size data, and using orthogonal contrasts.

	Sum of squares	d.f.	Mean square	F	P
Constant	3203.333	1	3203.333	372.802	.000
Contrast 1	20.000	1	20.000	2.328	.139
Contrast 2	166.667	1	166.667	19.397	.000
Within	232.000	27	8.593		

Table X.7.3 Anova table for equal cell size data, testing the entire comparison.

	Sum of squares	d.f.	Mean square	F	P
Constant	3203.333	1	3203.333	372.802	.000
Comparison	186.667	2	93.333	10.862	.000
Within	232.000	27	8.593		

2.8 RANDOM EFFECTS MODELS

Sometimes the levels of a factor are not fixed but arise from some sampling mechanism. Thus the groups might be different cities, sampled from the total population of cities in the country. In this case one will hope not merely to make inferences about the particular cities included, but will hope to generalize further to other cities. Situations of this kind are called *random effects* models.

In this book we concentrate on *fixed effects* models – cases in which one is concerned only with making statements about the factor levels actually included in the study. In our experience fixed effects models are the most common in the social and behavioural sciences. Nevertheless random effects models do occasionally arise. In a random effects model individual contrasts will not usually be of interest. It is only the general variation, the overall difference between category means which is normally of interest. This is, of course, simply the complete comparison on the groups. Thus the attributable sum of squares for a factor in a random effects model can be found by using the complete comparison for that factor and getting the computer to generate the associated attributable sum of squares.

The other major difference between fixed and random effects models arises during testing. Since the levels of the factor are obtained by a random sampling mechanism, the additional variance associated with this mechanism must be allowed for. The simple within-groups variation may be inadequate as an error sum of squares. Details of appropriate error terms will be found in texts on univariate analysis of variance, such as Winer (1971).

29

2.9 THE OVERALL MEAN LEVEL

It is appropriate here to say a word about the overall level of the response variable, that is, about the grand mean across all groups. This could be formally described using the 'contrast'

16(i)   1   1   1   1

in a four-group study although, of course, strictly this is not a proper contrast since its elements do not sum to zero.

In most studies the concern is only with differences between groups (i.e. with proper contrasts). For example, central interest normally lies in comparing different treatments, different experimental conditions, or different kinds of subjects. Very occasionally, however, the overall mean of the response variable is of interest and can be studied. An example of such a variable might be IQ where, having explored the differences between the mean IQs in the groups one might well wish to know the overall average IQ.

There are two circumstances in which we could legitimately estimate and test this overall mean. The first is when no between-groups differences are found. In this case one can simply regard the data as arising from a single group. If between-groups differences are found, however, then the overall mean is a function of group size – taking a larger sample from the group with the lowest mean will pull the overall average down. This implies that the value of the overall mean may just be an artefact of the experimental design.

The second set of circumstances holds whether group differences are found or not. If the group sizes in the sample are representative of the population's natural group sizes, then the overall sample mean estimates the overall population mean. It would then be legitimate to use this mean value, even if within the population there were differing values for each group.

In Section 2.3 we remarked that a complete comparison on g groups could be stated in terms of $(g - 1)$ contrasts. We should note here that the g values given by the value of the overall mean and these $(g - 1)$ contrasts are in all respect equivalent to the g group means. Either set of numbers can be obtained from the other.

2.10 MULTI-FACTOR DESIGNS

So far our discussion has been in terms of one factor. Now we turn to consider designs involving two or, indeed, any number of factors. Each of the factors classifies the subjects in different ways. To illustrate, let us return to the experiment described in Example 2 of Section 2.2. This has four groups: three different treatment groups and a control group. (We shall refer to this factor as the 'experimental' factor.) Suppose now we find on closer examination of the data that 90% of the control group are female but that only 20% of each of the three treatment groups are female. How might this affect our conclusions about the efficacy of the treatments?

It could, in fact, completely invalidate them. Consider the situation summarized in Table 2.1. This shows an experiment with the male/female percentages as above. The sample sizes in the groups are given in brackets. Each of the treatments and the control is assumed to have an identical effect on the females, producing a response of +1.0. Similarly, each of the treatments and the control is assumed to have an identical effect on the males, different from that of the females, producing a response of −1.0. The table shows these average within-group responses of +1.0 and −1.0 for females and males, respectively. Now, however, suppose that we had not taken account of sex. Then we would just have four cells, each with 100 subjects, producing the means given in the

Table 2.1 A sex-by-experimental group cross-classification. Each female is assumed to have an identical response of +1 and each male an identical response of −1. The mean responses in each group are given and, in brackets, the sample size in each group. The overall average for the four experimental groups is given on the bottom line but one.

	Tr.1	Tr.2	Tr.3	Control
Females	1.0 (20)	1.0 (20)	1.0 (20)	1.0 (90)
Males	−1.0 (80)	−1.0 (80)	−1.0 (80)	−1.0 (10)
Overall average	−0.6 (100)	−0.6 (100)	−0.6 (100)	0.8 (100)

last-but-one line. This shows an apparent extremely large difference (from −0.6 to +0.8) between treatment and control groups.

It is clear that if we expect the response to differ between sexes (i.e. if we expect there to be a sex effect) then we must somehow take this into account. In the present case we can do this by working with the two (sex) by four (experimental factor) cross-classification of the subjects into eight cells instead of into just four cells. Then, by looking at appropriate contrasts, as we shall do below, it is possible to examine the question: is there an experimental factor effect, over and above anything arising solely due to the sex effect and the inequitable distribution of the sexes within experimental groups? That is, we will be exploring the effect of the experimental factor *controlling for* the sex factor.

It will be instructive if we pause here to consider the two kinds of question that might be asked of the experimental effect in more detail. Firstly, we can do what we implicitly did in Section 2.2; that is, ignore sex and simply ask if any differences exist between the treatment and control categories. This is the simple four-group analysis, producing results as in the bottom lines of Table 2.1. Secondly, we can ask, as we have above, whether there are any differences between the treatment and control groups which cannot be explained by the sex factor. These questions are clearly different, and either may be the appropriate one – it depends on the researcher's objectives. A case when the first, simpler kind of analysis would be appropriate is the following. Suppose limitations on financial resources were such that a new hospital could be built in only one of four candidate towns (these four towns replacing the four levels of our experimental factor in the example). Then we might compare the towns to see in which the need was greatest, perhaps giving medical screening examinations to a random sample of the populations in each town. Clearly here we are interested in the assessment of the towns as such, regardless of any differences between types of people within each town, be they different on account of sex, age or any other factor.

More often, however, in the behavioural sciences one is really trying to address the question: for an individual what will the treatment effect be? Clearly in grouping subjects one wants to avoid grouping together subjects who behave differently, for precisely the reasons demonstrated in the example above, and this leads us to use the more complex analysis. A further important aspect of this distinction is that the numbers of subjects in the cells are often manipulated by the experimenter. We have already seen some of the advantages of having equal numbers of subjects. Obviously simple averages over artificially manipulated numbers of subjects will have little meaning and such a situation will preclude the use of the simpler analysis.

Thus, while both questions have meaning and may be relevant, it is our experience that the second kind, where effects are explored *controlling for* other factors, is the more common in the behavioural sciences. Note that we can put this notion of 'controlling for the sex effect' the other way round and ask: is sex alone *adequate* to explain the observed treatment effect? This is sometimes a useful way of looking at things, to which we will return in later chapters.

We have described here the idea of controlling for other factors from a particular perspective which we feel is appropriate for the behavioural and social sciences. A second motivation also exists, however, which we shall now briefly describe. In the above example, one might suppose that if the sex distributions were identical at each level of the experimental factor then there would be no advantage in adopting the more complicated two-factor design. However, this need not be the case. The presence of different types of subjects (men and women) in each level of the treatment factor means that there will usually be greater variability in the responses within each treatment level than if there were only one type. The differences between the observations within each treatment group will be inflated by any differences between responses in the different levels of the sex factor. Controlling for this by using sex as a second factor will usually

interaction. Such restrictions mean that better estimates of population means than the simple sample means may be available. The situation is precisely as in Section 2.4, and a computer program will derive the better (and more complicated) estimators automatically. It follows that better estimates of the sizes of differences between populations may be available than by applying the contrast elements directly to the sample means.

Again we note that if the assumption that certain patterns of differences are zero is incorrect, then the analysis can go badly wrong. We said in Section 2.4 that we would illustrate this point and we shall do so now for a two-factor example. The patterns of differences we will assume to be zero in this instance are the interaction effects.

■ EXAMPLE 8   This illustration is based on an example in Urquhart and Weeks (1978). It involves a two-factor (three-by-four) design with interest focused on estimating the difference between levels 1 and 3 of the first factor. We shall use an unsaturated model formed by ignoring the interaction effects. Thus we have overall in the cross-classification a comparison consisting of contrasts representing only the complete comparison for each of the main effects, but not their interactions. Thus we suppose (rightly or wrongly) that there are no interactions.

We shall consider four cases according to whether the cells have equal or unequal numbers of subjects and according to whether the no-interaction assumption is correct or incorrect. Table 2.4(a, b) shows the distribution of the cell sizes for the equal and unequal cases respectively. Table 2.4(c, d) shows the associated respective patterns of weights which should be applied to the sample means to obtain efficient estimators of the single contrast representing the (level 1–level 3) difference, when one assumes no interactions exist. These optimum

Table 2.4 Numbers of subjects in each cell: (a) equal case, (b) unequal case. The most efficient estimators of the (level 1–level 3) difference, assuming no interaction case, are given by applying the weights in (c) and (d) to the cell means responses in each cell.

(a)				(b)			
3	3	3	3	1	5	1	1
3	3	3	3	1	1	7	3
3	3	3	3	7	3	1	5

(c)				(d)			
.25	.25	.25	.25	.198	.474	.147	.181
0	0	0	0	.043	−.060	−.061	−.078
−.25	−.25	−.25	−.25	−.241	−.414	−.086	−.259

weights, obtained using a computer program, have been written for convenience as a block (rather than on a single line as in previous sections). The differences between the optimum weight patterns are striking, but what is particularly noteworthy is that (d) includes non-zero weights for level 2 of the first factor. This means that even though our question does not relate directly to level 2, the restrictions imposed by assuming there to be no interactions mean that better estimators (in terms of smaller variance) result if the level 2 means contribute to the estimator. Note that these weights will apply to any experiment with the given cell sizes, whatever the observed cell means in them might actually be.

Table 2.5 (a) Group means for a no-interaction case; (b) Group means when an interaction is present.

(a)				(b)			
1.0	2.0	3.0	4.0	0.0	10.0	0.0	0.0
3.0	4.0	5.0	6.0	0.0	0.0	0.0	0.0
5.0	6.0	7.0	8.0	0.0	0.0	10.0	0.0

We now consider two possible situations in which these two experimental designs could be applied. Table 2.5(a) gives the underlying population means for each cell in a case where the no-interaction assumption is correct. Whether we have a design with equal cell numbers (2.4(a)) or unequal numbers (2.4(b)), were the appropriate associated set of weights from Table 2.4 to be applied to means reflecting this pattern in Table 2.5(a), it would yield the estimate of the (level 1–level 3) effect to be -4.0.

Table 2.5(b) gives the underlying population means for a case when interaction is present; that is, the assumption of no interaction is incorrect. In the case of equal cell sizes (2.4(a)), the associated weight pattern in 2.4(c) gives the estimate of the (level 1–level 3) effect as 0.0. In the case of unequal cell sizes (2.4(b)) the associated pattern of weights given in 2.4(d) gives the estimate as 3.9. We see that in making an incorrect assumption of no interaction we have been led to an incorrect result in this second case, where groups have unequal sizes. Thus, as we said in Section 2.4, one must be confident that patterns really are zero before dropping them from the comparison. ■

We should remark here that in any case if an interaction is present then an estimate of the main effects may be of limited value. If, for example, response increases with increasing dose for males but decreases for females, of what value is it to say that on average (the main effect of dose) there is a no dose effect at all? This point will recur, especially in within-subjects designs in later chapters.

Exactly the same point arises in single-factor designs. Suppose that an experiment uses four equally spaced dose levels of a drug, making a single experimental factor. Supose further that, as in Section 2.6, Example 7, we are interested primarily in the rate at which response changes with increasing dose (the contrasts of set 15 represent this information). If the rate of change from level to level gets greater the larger the dose, then of what value is the overall average rate of response? This average depends on the range of levels chosen for the experiment. It is for this reason that in Example 7 the sequence of tests employed ensured that the highest-order rates of change were in fact zero before estimating such an average.

Finally in this section, we return to the question of controlling for other factors, introduced in Section 2.10, in order to make some general points. In that section we showed how the estimated effect of a factor depended on whether or not we controlled for others, and that the appropriate way to proceed depended on the precise research questions. Here we shall generalize those ideas. We shall illustrate with a two-by-two design involving treatment (levels A and B) by dose (low and high). To keep things simple we shall suppose that treatment and dose have no interaction.

If we arrange the four cells in order A then B at low dose, then A then B at high dose, the contrasts

22(i)	1	-1	1	-1
22(ii)	1	1	-1	-1

describe, respectively, the treatment (A–B) effect and the dose effect. If the four cells have equal sample sizes then we have an especially simple case where these contrasts are orthogonal: so each can be separately estimated by applying the contrast elements directly to the cell means.

Suppose now, however, we take the general case where the cell sizes are unequal. This means that the contrasts in set 22 are no longer orthogonal, so that the estimated size of one depends on whether or not the other is simultaneously estimated or alternatively assumed to be zero. In other words, the estimated size of one depends on whether or not we control for the other. In terms of the factors, we have that the estimated treatment effect depends on whether we control for dose, and the estimated dose effect depends on whether we control for treatment.

To control or not to control? As discussed in Section 2.10, the answer must depend on the researcher's aims. This situation, with the two kinds of questions, is perfectly general and presents a central question that the researcher must answer. The fact that for orthogonal contrasts the two questions are the same is the principal advantage of orthogonal contrasts. However, the fact that it means one can avoid thinking carefully about exactly what one wishes to know does not justify changing the research questions because they happen to correspond to non-orthogonal contrasts.

39

2.13 TESTING IN MULTI-FACTOR DESIGNS

In Section 2.6 we described the process of testing the value of a single pattern of differences. This pattern may have been part of a complete comparison, so that the contrast elements were applied directly to the sample means, or the pattern may have involved more-complicated weighting procedures to derive the estimate. In either case, exactly the same procedures and tests apply in a multi-factor design as in the single-factor case.

Similarly, the process of testing comparisons, complete or incomplete, as described in Section 2.7, also applies to the multi-factor case. In particular, we also have the choice of whether or not to pool error variation and within-group variance for the denominator of the F-test.

In the multi-factor case we must consider which factors to control for. As discussed above, the answer depends on precisely what questions one wants to address. In Section 2.6 we gave an example where the contrasts, measuring rates of change, fell into a natural order. It was appropriate to start with a complete comparison consisting of all the contrasts (the saturated model) and test each contrast sequentially. If it proved not to be significantly different from zero, it was omitted from the comparison; and so on until one was found to have a non-zero value.

The same principle applies in multi-factor designs. In particular, as we have already noted, main effect contrasts make limited sense if interactions are non-zero; and these in turn make little sense if higher-order interactions are non-zero. Thus a natural order is induced and it follows then that the highest-order interactions must be tested first, as part of a complete comparison. In designs with more than two factors one begins with the highest-order interaction and works one's way down, stopping when a non-zero interaction or main effect is discovered. It should also be noted that, generally, the estimates remaining in a comparison must be recalculated after deciding that some contrast is zero. The reasons for this have been described at length in Section 2.4.

For contrasts referring to different (higher- or lower-) order effects and main effects, the sequence of testing is clear. But what about contrasts of the same order – for example, the main effects associated with different factors? Should we or should we not control for one when testing for the other? This is, of course, precisely the question in Section 2.12. And the answer, as there, is . . . it depends on what one's question is.

The sum of squares for a factor produced by controlling for all other factors is called the *unique sum of squares* attributable to that factor. That produced in a sequential way, controlling for some contrasts and assuming others higher up the sequence to be zero, in a cumulative manner, is called a *sequential sum of squares*; its value depends, obviously, on which contrasts are controlled for, and which assumed zero at that point in the sequence.

Display 8 A two-factor example

Still restricting ourselves to the single variable M1, a third researcher suspects that there might be a sex difference in the responses, and therefore took note of the subjects' sex when recording the other data. The data are shown in Table X.8.1. As before, the primary aim is to see if there are differences between the different levels of the condition factor. The researcher therefore conducts an analysis of the two-way design with sex and condition as factors. The steps carried out in running computer analyses and interpreting the output are as follows.

First, those contrasts relating to interaction are examined, and grouped together to give a single overall test of the interaction comparison since, at this stage anyway, the individual interaction contrasts are not of interest. Table X.8.2 shows that this interaction contrast is not significant. It is therefore reasonable to assume that the two sexes have the same pattern of differences between means. This does not mean that the sexes behave similarly in the condition groups – females might score consistently higher than males – but merely that the *patterns* across condition groups are the same for the two sexes (if males do best in the lectures group, so also do females). The interaction contrasts are dropped from the overall comparison and work proceeds with the resulting incomplete comparison.

The next step is to explore the condition contrasts of interest while including sex in the comparison (but having dropped sex-by-condition interactions). Suppose we are concerned with the contrasts of Display 2:

$$
\begin{array}{ccc}
1 & 0 & -1 \\
0 & 1 & -1
\end{array}
$$

The results appear in Table X.8.2. When estimating and testing these contrasts both of them are kept in the comparison – though if one were to be found to be zero it could arguably be dropped when studying the other, since the results would be more efficient (see text). Again we remark that all terms (other than interaction) in Table X.8.2 are tested without including interaction in the comparison.

On closer examination of Table X.8.2 the researcher notices that the sex effect is in fact not significant. On the assumption that this lack of significance means that the sex effect is zero, then more efficient estimates will be obtained if sex is dropped, and the incomplete comparison is used just involving the two condition contrasts.

Table X.8.1 The data for Display 8.

G1:	Males:	8, 9, 16
	Females:	11, 13, 9
G2:	Males:	13, 16, 9, 11, 16
	Females:	13, 16, 12, 13, 16
G3:	Males:	6, 6, 6, 10
	Females:	6, 9, 4, 5, 6, 11, 8, 9, 9, 3

Between-subjects analysis

Table X.8.2 Anova results for Display 8.

	Sum of squares	d.f.	Mean square	F	P
Constant	2863.568	1	2863.568	404.268	.000
Sex	0.911	1	0.911	0.129	.723
Contrast 1	67.940	1	67.940	9.592	.005
Contrast 2	242.929	1	242.929	34.296	.000
Sex-by-condition	1.589	2	0.794	0.112	.894
Within	170.000	24	7.083		

However, as explained in the text, if this assumption is wrong then bias is introduced. Since the researcher originally suspected there was a sex effect it might be as well to retain it. Analogous arguments apply to whether or not to drop the interaction contrasts from the comparison.

2.14 NESTED DESIGNS

A special case of the general multi-factor design is one in which the factors are nested, for example, when people are clustered into cohorts and the cohorts into treatment groups. We have here two factors, 'cohorts' and 'treatments', and the cohorts are subdivisions of the categories of the treatments factor. We thus describe the cohorts factor as *nested within* the treatments factor. Unlike a general cross-classification, each level of the cohorts factor occurs in only one level of the treatments factor, and not in all, or even several, of them.

Thus with nested designs there is an automatically enforced ordering to the factors which we must take into account when testing for their effects. In particular, it is pointless controlling for the nested factor (in this case, cohorts) when assessing the nesting factor (here, treatment groups) because this will eliminate the effect of the nesting factor, treatment groups: there are no differences between treatments which are not also differences between cohorts. Thus in a nested design it is appropriate only to control for the more inclusive factor when assessing differences due to subdivisions. The reverse procedure is not sensible.

Note, too, that nested designs almost always use a random effects model and that it is therefore necessary to use an error sum of squares different from the usual within-cells error.

2.15 SOME IMPORTANT COMPARISONS

In this section we give a brief description of some of the more important ways in which comparisons can be described. In all the examples given below the set of summary contrasts is a complete set – any pattern of between-groups differences may be derived from the set. These examples are important because each represents a set of summarizing contrasts in which each contrast itself answers a useful question about the differences between group means. Each contrast set applies to the levels of a single factor and, if desired, each factor in the design may be described using a different summary set of contrasts, as in Section 2.11. The comparisons are illustrated below only on a factor involving four groups in order to avoid detailed algebraic notation. The extension to factors with a different number of groups should in each case be self-evident.

1. A common aim in research is to compare each of several distinct groups with one other group. For example, in a drug trial in which each of several drugs is being compared with a placebo control. If the first group is the control group, then the three suitable contrasts are:

1	−1	0	0
1	0	−1	0
1	0	0	−1

2. Most elementary analysis of variance texts take as the fundamental set of contrasts the set which compares each group in turn with the average of the other groups. That is

3	−1	−1	−1
−1	3	−1	−1
−1	−1	3	−1

A final contrast $(-1 \ -1 \ -1 \ 3)$ is superfluous since it can be computed from the complete comparison defined by these three.

3. In many cases there is some intrinsic order to the groups defined by a factor, as in Example 7 in Section 2.6. Perhaps they represent increasing doses of a drug treatment, or perhaps increasing levels of educational attainment. In such cases it is sometimes the successive differences between the groups which are of particular interest. Thus each group is compared with the immediately following group. The relevant contrasts are

1	−1	0	0
0	1	−1	0
0	0	1	−1

4. One of the most common orthogonal sets of contrasts is the Helmert system. In this each group is compared with the average of the following groups. Note again that some kind of ordering for the groups is implied. The relevant contrasts are

3	−1	−1	−1
0	2	−1	−1
0	0	1	−1

5. Another important orthogonal set used in certain contexts is the polynomial system, which describes the differences between groups in terms of the trends across them (see Example 7). The first represents a linear trend or slope across groups; the second is the quadratic trend or curvature; the third is the cubic trend, or rate of change of curvature; etc. These contrasts are most appropriate when there is a numerical characteristic of the groups. For instance, the groups may represent different dosage levels of a drug, the number of days since the beginning of the experiment, or different temperatures, and so on. The orthogonal polynomial contrasts shown here are the simplest form, which occur when the intervals between the levels of the groups are equal. Some programs will generate contrast weights for unequal spacings.

-3	-1	1	3
1	-1	-1	1
-1	3	-3	1

2.16 FURTHER READING

Any text on univariate analysis of variance or experimental design will provide background reading for this chapter. Notable ones are Keppel (1973), Winer (1971), Cochran and Cox (1957), Kempthorne (1952), Scheffe (1959) and Rouanet and Lépine (1977).

Papers which describe the approach to analysis of variance via direct estimation of cell means, rather than via the linear model form, are Urquhart and Weeks (1978) and Hocking and Speed (1975).

Within-subjects ———————— 3
analysis

3.1 INTRODUCTION

In Chapter 2 we considered differences between distinct groups of subjects. That is, the research questions we examined were related to patterns of differences between groups of subjects: between-subjects analysis. Such questions apply equally to both univariate and multivariate analysis of variance. In this chapter we turn to the uniquely multivariate aspect, namely an examination of the several distinct response variables measured on each subject, and the relationships between them. By implication from this, since the variables are measured on each subject, such questions relate to *within-subjects analysis*. Alternatively, since it is with patterns of difference between such variables that we will be concerned, we will sometimes call it *between-variables analysis*.

Formally, the extension from a univariate to a multivariate design is straightforward. Instead of subjects being characterized by their score on a single variable, they are characterized by their scores on several variables – that is, by a vector of scores. The immediate implication of this, and what makes multivariate analysis of variance exciting, is that each of the between-group contrasts discussed in the preceding chapter now takes on a vector of values, one for each of the variables in the analysis. Thus, whereas in the univariate case the estimated value of a contrast is a single number, in the multivariate case its value depends on which variable one considers, or on which combination of variables, or on which set of variables one examines simultaneously.

A second complication following from the use of multiple variables is that we will no longer have a single sum of squares in our tests. At the very least one would expect one such a sum for each of the variables and in fact we also have to take into account the sums of products of different variables. We can think of this set of sums of squares and products as a matrix with its rows and columns indexed by the variables. The (ij)th element of this matrix will be the sum of products relating to the (i)th and (j)th variables. Its (i)th diagonal term (when i equals j) will be the sum of squares for the (i)th variable that would be used in a univariate analysis. We shall call such generalizations of the sum of squares a *sum of products matrix*. It is clear from this definition that a sum of products matrix is symmetric (i.e. its (ij)th element equals its (ji)th element). What is also true is that if one divides each element of such a matrix by the appropriate constant number, one obtains a covariance matrix between the variables, with the

45

diagonal entries being the variances of the variables. We shall see in Chapter 4 that in statistical tests these sums of products matrices play a role in multivariate analysis that is directly analogous to that of the straightforward sum of squares in univariate analysis.

Having established this basic parallel, we turn in the next section to the sorts of questions that a multivariate analysis can ask of the data. First, however, note that we shall assume that each subject has a complete vector of observations. If, for one reason or another, missing values occur, it will be necessary to handle the analysis by approximate methods. These will usually involve prior estimation of the values which are missing and then proceeding as though we had a complete set of data.

Display 9 A sum of products matrix

Consider the set of numbers $\{6, 7, 11, 3, 2\}$. The sum of squares of this set is

$(6^2 + 7^2 + 11^2 + 3^2 + 2^2) = 219$.

Now consider the set $\{1, 3, 10, 7, 7\}$. This has sum of squares

$(1^2 + 3^2 + 10^2 + 7^2 + 7^2) = 208$.

The sum of products of these two sets is

$(6 \times 1) + (7 \times 3) + (11 \times 10) + (3 \times 7) + (2 \times 7) = 172$.

We obtained this by multiplying each number in the first set by the matching number in the second, and adding. We would, of course, have obtained the same result if we had taken the second set first.

We can write these results in matrix form as

219	172
172	208

This is a sum of products matrix.

Usually each set corresponds to a single variable, containing one number for each subject. If there are ten subjects measured on four variables, for example, this means that each sum of squares or products will be made up of ten terms. There will be four such sums of squares, one for each variable, and six such sums of products, one for each pair of variables. The four sums of squares appear down the leading diagonal of the matrix (in the two-variable case above they are 219 and 208) while the six sums of products appear in the triangular array above the diagonal. They are repeated below (reflected in) the diagonal – where they correspond to the same pairs but with each pair taken in the reverse order from the upper triangle.

3.2 TYPES OF QUESTION

Chapter 1 outlined the sort of questions that we are likely to ask in a multivariate analysis. We recapitulate these types in this section and go into details in the following sections. At one extreme we can consider the data as a set of single variables and ask several questions of an essentially univariate nature. This would be appropriate in studies where each response variable was of interest in its own right. For example, if height, weight, and strength were measured in a nutrition study it is likely that one would wish to examine them separately. The aim would (probably) not be to discover if different groups of subjects differed overall, but to find out if (a) the groups differed in height, (b) the groups differed in weight, and (c) the groups differed in strength. We shall term such studies *multiple univariate* because this is what they are.

In between two extremes of approach lie questions of the type which we call *structured multivariate*. These are questions addressing specific relationships between variables and the interplay between these relationships and the different groups of subjects. An example of a structured multivariate study is a *repeated-measures design*, in which the several response variables are results of the same test (or measurement) applied at a number of different times. Here it is unlikely to be between-group differences at each individual time point which are of interest, but rather differences in the way groups develop over time. Such questions involve combinations of the responses at each of the time points, that is, combinations of the variables, usually in a specifically structured way.

Finally, at the opposite extreme from multiple univariate questions, we have *intrinsically multivariate questions*. Here we seek to address questions such as: does this set of variables as a whole indicate any between-group differences? For example, a questionnaire might consist of a number of items, none of which is of any great interest in its own right, but which must be combined somehow to yield a comparison between groups.

Although we have delineated three types of questions, in fact any actual study will probably contain a combination of the three types. For example, having identified that some group difference exists in an intrinsically multivariate problem, one might want to follow this by trying to identify the particular variables that are chiefly responsible for the difference. None the less, the tripartite typology is a useful one to bear in mind. It does help one state the research objectives more clearly. Having said which, of course, it must be said also that stating research objectives clearly is not always easy. An alternative formulation of the two extreme types of research question above might help:

- Multiple univariate questions are appropriate if one wishes to see if the groups differ on any of the particular, given variables. In statistical tests, one might want to consider an overall significance level for all variables simultaneously (as described in Chapter 5), but otherwise each variable is given separate consideration.

- Intrinsically multivariate analyses are appropriate if one wishes to see if there are *any* between-groups differences at all, when the only information available is the scores on the given variables. No prior structural information about the variables is considered.

47

3.3 MULTIPLE UNIVARIATE STUDIES

In studies of the 'multiple univariate' kind, it would not be appropriate to ask questions about general multivariate differences. The appropriate method of analysis in such studies is to carry out separate univariate investigations. However, there are difficulties associated with such an approach, and the reader must be aware of them. These difficulties are discussed more fully in Chapter 5; here we merely introduce them informally, before moving on to consider structured and intrinsic studies.

The basic problem with multiple analysis is that if one carries out a number of tests of true null hypotheses, each at a specified significance level, then the overall probability of incorrectly rejecting any one or more of the hypotheses is greater than that specified by the significance level. Indeed, depending on the levels of such component tests and the number of such tests, the overall level can be arbitrarily large. This point is the same one that leads us in univariate analysis to consider multiple comparison techniques. The dilemma introduced is thus as follows: if we control the error in each component test at a conventional level (say, the 5% level), then the overall probability of error is large. If, on the other hand, we restrict the overall probability to a conventional level, the individual levels of the component tests can be exceedingly small. The distinction here is between experiment-wise error rate (the overall significance level) and test-wise error rate (the significance level of each component test).

The dilemma, of course, is in choosing which level one should control. The answer, as always, must be 'it depends . . .'. In this case it depends on the sorts of risks of being wrong that the researcher is prepared to take.

3.4 STRUCTURED MULTIVARIATE STUDIES: A SIMPLE EXAMPLE

We noted above that in many multivariate studies the individual variables measured on each subject are of less intrinsic interest than are the relationships between them. A simple example of such a situation is when each subject is measured at two time points – perhaps before and after a treatment – and we wish to show whether there has been a change. Although the same response (say, heart rate) is measured at each time, since two distinct measures are obtained we can regard this as a multivariate (bivariate) problem. In this case the primary question of interest relates not to the individual measures but to the *change*, that is, to the difference between them. Thus, an analysis might be based on the difference score. (Note that it would then have been reduced to a univariate analysis.) The transformation from the measured variables to ones more relevant to the questions at issue is a vital concept in multivariate analysis, as we shall see below.

The above illustration relates to a single group and addresses the question: Has this group's mean changed from time 1 to time 2?' We extend this illustration to two groups, in order to show the wealth of questions that can be generated in a structured multivariate analysis. If two groups were involved and the analysis was based on the same derived time-change variable, then we could consider two questions. Firstly, by looking at the between-groups contrast, addressing the issue of whether there is a difference between the groups, we could consider the question: 'Do the two groups change in distinctly different ways from one time to the next?' This can also be viewed as asking if there is an 'interaction' between the time and group effects. The second question arises if we have no reason to expect a difference between groups on the derived time-change variable. Then we might merge the two groups and simply ask if their combined average on the time-change variable is zero. This has effectively led us back to a one-group question, and to asking: 'Is there a change over time?' That is, is there a main effect for the time factor?

So far we have examined the time-by-group interaction and the time main effect. It remains to consider the group main effect. Again, it would usually be legitimate to pose such a question only if we found no interaction, which implies that we expect the difference between the groups to be the same at time 1 as at time 2. The question 'is there a group main effect?' is asking whether this (presumed identical) group difference is zero. We can explore this question by calculating a new derived variable for each subject – the average (or sum) of that subject's scores at time 1 and time 2. This averaging improves the accuracy of the estimates and the power of the tests. Note that to use the average instead of the sum requires only a rescaling, which will not affect the test results.

We can summarize the procedures above as follows: for each subject the original two variables (measured at time 1 and time 2) are transformed to two alternative derived variables: the sum and the difference of the two time scores; these two new variables directly address the questions of interest.

We can write the difference between the two variables in the following way. If x_1 is the score at time 1 and x_2 is the score at time 2, the (new) difference variable is $(x_1 - x_2)$. This can be written as $(+1)x_1 + (-1)x_2$. That is, we multiply x_1 and x_2 by $+1$ and -1, respectively, and add the results. Thus the set of numbers $(+1 \ -1)$ describes the pattern of differences between the original variables. The reader will, of course, recognize instantly that this is simply a contrast – only now we are applying it to the variables rather than to the groups. This, then, is our first hint at the general power of the contrast approach. We shall develop it in the context of *between-variables* questions below.

Note that the particular contrast we were interested in – that showing a simple difference between the two variables – was pre-specified. It was an explicit part of our research question, as was the other new derived variable, the sum of the two variables. This ability to pre-specify the transformation we want, since it involves additions and subtractions of the variables, will usually require some sort of commensuration across the variables to retain meaning in these calculations (it is of no use to add pounds to inches, unless we are very sure of what we intend to do with the result). This commensuration will usually be the case when the same variable is measured repeatedly, or when the scale of measurement is the same (a percentage or proportion could be an example).

The example given here is particularly useful for expositional purposes because it illustrates simultaneously two types of structure. It can be considered as a one-way classification of the variables – a factor of 'time' at two levels, where the difference between the two variables represents the effects of time. Alternatively, given it concerns change over time, it can be thought of as a simple repeated measures analysis, involving only two time points, and requiring the examination of a linear trend. Each of these formulations will be extended to more-complex designs below.

Display 10 A structured bivariate example

In Chapter 2 all our examples were univariate. Now we extend to the multivariate case. In particular, this first example will consider a structured bivariate study.

A researcher is interested in the three condition groups described in Display 1, but also in how time pressure might influence the results. To explore this each subject is tested under two different regimes: one when there is no time pressure, and the other when the subjects were rewarded if they finished quickly. The same test is administered in each case, but since it is recorded twice we regard the experiment as producing two distinct variables. These will be labelled M1 (when there was no time pressure) and M2 (when there was time pressure). The data are given in Table X.10.1.

Table X.10.1 Data for Display 10.

Group 1		Group 2		Group 3	
M1	*M2*	*M1*	*M2*	*M1*	*M2*
8	8	9	8	6	11
7	4	9	7	6	1
12	4	9	7	6	0
11	7	10	4	10	7
15	10	8	4	6	1
13	9	10	7	9	4
		15	10	4	1
		15	9	5	1
		9	6	6	2
		16	8	7	3
				7	10
				11	6
				12	5
				3	4

Table X.10.2 Anova results for difference of M1 and M2 scores.

	Sum of squares	d.f.	Mean square	F	P
Constant	357.887	1	357.887	40.601	.000
Contrast 1	4.200	1	4.200	0.476	.496
Contrast 2	5.833	1	5.833	0.662	.423
Within	238.000	27	8.815		

Considering comparisons between groups, the researcher is again concerned with the two contrasts

```
1        0       -1
0        1       -1
```

Neither of these is expected to be zero a priori, so both will be included in a comparison to test each of them.

The first aim is to see if the time pressure causes a difference in the responses. That is, we are concerned with the difference between M1 and M2. A new score is thus calculated for each subject, the difference of M1 and M2, and this is analysed. The anova table is given in Table X.10.2.

This table could be obtained by computing the difference score prior to entering the manova program, and then analysing the resulting single variable. Alternatively, most manova packages permit one simply to specify what weighted sum of the variables is needed and they calculate it automatically. The resulting weighted sum might have a different scaling from the pre-computed version but as explained in the text this does not affect test results (it only affects sizes and sums of squares and products). Also, of course, if one pre-computes the variable then the program regards it as a raw variable, whereas if the program derives it, it can recognize it as a derived variable. This can affect labelling in the anova table.

If we first consider the test on contrast 1 in Table X.10.2 we see that there is no significant difference between groups 1 and 3, as measured on the (M1−M2) difference score. This can be expressed in several different ways:

- any difference between groups G1 and G3 on M1 is the same as the difference between groups G1 and G3 on M2.
- any difference between M1 and M2 in G1 is the same as the difference between M1 and M2 in G3.
- there is no interaction between the two-level conditioning factor, with levels time pressure and no time pressure, and the factor consisting of groups G1 and G3.

We also note from Table X.10.2 a highly significant constant. This is measuring the overall effect of (M1−M2), suggesting that although M1 and M2 do not differ in different ways between the two groups, there is some constant difference between them.

Having seen that, if M1 and M2 differ at all, then they differ in the same way between groups G1 and G3, the experimenter can now go on to see if they differ at

Within-subjects analysis

Table X.10.3 Anova results for average of M1 and M2 scores.

	Sum of squares	d.f.	Mean square	F	P
Constant	3266.831	1	3266.831	274.780	.000
Contrast 1	102.900	1	102.900	8.655	.007
Contrast 2	142.917	1	142.917	12.021	.002
Within	321.000	27	11.889		

Table X.10.4 Analysing a comparison using the (M1–M2) difference.

	Sum of squares	d.f.	Mean square	F	P
Constant	178.944	1	178.944	40.601	.000
Comparison	3.733	2	1.867	.424	.659
Within	119.000	27	4.407		

all. Since they differ in the same way, this could be done by examining either one of them. Or a more powerful test could be obtained by working with their average (or sum, since this just represents a rescaling, which will not affect the tests). The results are given in Table X.10.3. The difference between G1 and G3 on this derived variable is seen to be highly significant.

Tables X.10.2 and X.10.3 show that contrast 2 has the same pattern of results. Thus we can summarize by saying that, while time pressure seems not to affect the response difference between groups G1 and G3 and between G2 and G3, there are non-zero response differences.

By way of a further example, let us suppose that the researcher is simply interested in whether there are any between-groups differences in response to time pressure. The concern is not with any particular contrasts, but only with the overall comparison. Again the response to time pressure will be measured by (M1−M2)/2. The test results are shown in Table X.10.4.

This Display has shown how structured multivariate problems can be reduced to a sequence of simple univariate analyses. More generally, they reduce to a sequence of multiple univariate and intrinsically multivariate problems. We shall see further examples of this after considering intrinsic multivariate analyses.

3.5 REPEATED-MEASURES STRUCTURES: TREND ANALYSIS

A natural extension of the simple two-time-point design discussed in the preceding section occurs when each subject is measured at several time points, for example immediately after, one month after, two months after, and three months after treatment. Again the actual observed values will usually be of less interest than simple derived variables indicating patterns of change over time

Thus a researcher may be interested in whether there is a uniform improvement over time or whether the rate of improvement falls off as time progresses. If more than one group is involved, questions might relate to comparing patterns of change over time across groups.

Questions involving patterns of change over time can be tackled using polynomial transformations, as outlined in the context of between-group contrasts in Chapter 2. The difference is that whereas there the contrast elements were related to group means, here they relate to variables (just as the $+1$ and -1 in Section 3.4 referred to the variables measured at times 1 and 2).

Suppose now that the original variables are uncorrelated. Then it can be shown that if orthogonal contrasts are used, the new derived variables will also be uncorrelated. Note also that since each subject has a single score on each variable (remembering that the same thing measured at two times counts as two variables), it follows that for the within-subjects design we can use the simpler criterion of orthogonality given in Section 2.5 (ignoring the quantities n_i, which are all effectively equal to 1). Of course, the original variables will typically be correlated, but we note that by using orthogonal contrasts no 'spurious' correlation is induced by the transformation process.

If, in addition to the variables derived by applying polynomial contrasts to the raw variables we include the overall average score over time, we obtain the set of derived variables:

(a) the overall average across time;
(b) the linear trend across time (i.e. slope, or rate at which the response changes with time);
(c) the quadratic trend across time (i.e. the rate at which the slope changes);

and so on.

Note that (a) is often replaced by the total across time (i.e. summing each variable over the times at which it is measured). This makes no difference to tests of significance.

Transformations like this take the original measures (recorded at several time points, or doses, or levels of noise, etc.) and produce an equal number of alternative variables. The simple two-time-point case explored above has only components (a) and (b) of these (it starts with two variables and finishes with two). If four time points (or doses or whatever) were involved, then the most complicated component would be a cubic (3rd order). Characterizations of change other than the polynomial contrasts may be desired – perhaps the Helmert system described in Section 2.15 would be useful – and indeed any specified transformation can be used in an identical way.

Questions relating to patterns of change over time (or dose, etc.) are studied by examining all except the first (the overall average) component. For a particular pattern of change, say a linear trend, we could examine the appropriate trend variable, but for general patterns of change we would need to look at

Within-subjects analysis

all of them. Thus, for example, to answer the simple question 'do the groups behave in any ways differently over time?' we could perform an intrinsically multivariate analysis using the linear, quadratic, etc., components simultaneously.

Display 11 A trend analysis

One researcher was concerned only with groups G1 and G2 and wanted to characterize the relative way that performance degraded (or maybe even improved) after the lectures or formal practice had ceased. The measurement of problem-solving skill M1 was therefore repeated at three times: immediately after the lectures or practice, one month after, and two months after.

We can describe the aim as being to describe the way the between-groups difference changed over time. Put yet another way, the aim was to describe the difference between patterns of change over time. This last way makes it clear how we can approach the problem: first we describe the patterns of change over time in each group and then we compare these patterns.

A common way of describing patterns of change over time is through the use of derived variables which are polynomial components of the original scores. In this case, with three time points, we will be able to produce only two derived polynomial components. These are the linear slope, obtained by applying the weights

$$1 \qquad 0 \qquad -1$$

to the three variables in the order given above, and the quadratic slope, obtained using

$$1 \qquad -2 \qquad 1$$

Consider an example. Suppose that a particular subject scores 15.0 immediately after the lectures, 13.0 one month later, and 11.0 two months later. Then the derived linear and quadratic components are, respectively

$$(15 \times 1) + (13 \times 0) + (11 \times (-1)) = 4$$

and

$$(15 \times 1) + (13 \times (-2)) + (11 \times 1) = 0.$$

The fact that the quadratic component is zero just tells us that the deterioration in performance over the second month is the same as that over the first month. From this we can see that each subject yields a pair of derived variables. Subsequent between-groups analysis will be based on the group means of these derived variables.

Of course, there is also a third component which can be derived from three scores – a component proportional to the mean of the three, obtained by supplying the weights

$$1 \qquad 1 \qquad 1$$

Table X.11.1 Data for simple two-group trend analysis. M1 is measured at times 1, 2 and 3. The derived variables are also given.

	Raw variables			Derived variables		
	Time 1	Time 2	Time 3	Sum	Linear	Quadratic
G1						
	11	13	4	28	7	−11
	10	16	6	32	4	−16
	16	15	0	31	16	−14
	11	14	5	30	6	−12
	9	17	7	33	2	−18
	3	15	8	26	−5	−19
G2						
	21	33	33	77	−2	−22
	21	16	22	59	−1	11
	18	30	22	70	−4	−20
	20	37	25	82	−5	−29

This, of course, just gives a measure of the average level or score of this subject. It does not contain information about the pattern of change over time.

Table X.11.1 shows the scores for the two groups of subjects that this experimenter used. This table also shows the three derived variables obtained by applying the above sets of weights. Note that sums of squares reported in anova tables may differ from those produced by using these derived variables. This is because any sets of weights proportional to the above can be used (any rescaling is permitted). The test results remain the same because both between- and within-groups sums of squares will be subjected to the same rescaling.

We can analyse the results as a sequence of univariate questions. First we look to see if there is a quadratic component difference between the groups. If there is, then our description includes both quadratic and linear components. If there is not, then we can drop the quadratic component and examine the linear component. Table X.11.2 shows these tests. Note that each derived variable has its own error sum of squares. The quadratic component is not at all significant, so we can readily drop this. The linear component is almost significant at the 5% level, so we might hesitate about dropping this (but see the discussion of simultaneous test procedures later – perhaps the fact that we are performing several tests should cause us to take a stricter significance level). If we retain it then our analysis stops here – an adequate description of the time trend needs both linear and mean polynomial components. If we drop it, then we see that the mean component is highly significant.

Another researcher looks at the same data with a more specific question in mind. This one simply wants to know if any between-groups difference which may exist is constant over time. If the answer is yes, it is constant, then a further question is whether this constant difference is in fact zero.

The first of these questions is not specifically aimed at describing patterns of change over time. Any descriptive form would do – we merely need to see if there

Table X.11.2 Anova table for the difference in time evolution of the two condition groups.

	Sum of squares	d.f.	Mean square	F	P
Quadratic					
Condition	.000	1	.000	.000	1.000
Within	166.333	8			
Linear					
Condition	76.800	1	76.800	4.995	.056
Within	123.000	8			
Mean					
Condition	1411.200	1	1411.200	102.014	.000
Within	110.667	8	13.833		

is no change in the difference score. The polynomial description above could be used perfectly well to answer these questions. But so also could other descriptions. For example, we can describe the pattern of change over time by the change from time 1 to time 2, given by using the weights

$$1 \qquad -1 \qquad 0$$

followed by the change from time 2 to time 3, given by

$$0 \qquad 1 \qquad -1$$

A simultaneous test on the derived variables using these two sets of weights will tell us if there is any change over time. Again, however, this is an intrinsically multivariate test, to which we return below.

3.6 REPEATED MEASURES BY UNIVARIATE METHODS

Readers familiar with univariate analysis of variance may recollect that repeated-measures studies such as that just described can also be tackled using univariate techniques. To do this one basically proceeds as if the repeated factor ('time' or whatever) were a between-subjects factor, and treats 'subjects' as comprising a random effects factor (see Section 2.8). Univariate approaches, however, are not in general valid. They are only valid if conservative tests are adopted or if certain restrictive assumptions are true.

Taking the former first, a valid univariate test of the complete comparison for the repeated factor results if, in the *F*-test, the degrees of freedom of both the numerator and denominator mean squares are divided by the number of degrees of freedom for the repeated factor. In the case of a repeated factor with t levels (e.g. time, with observations being made at t time-points), this divisor will be $(t-1)$. However, this test is likely to reject a true null hypothesis less often than the nominal significance level would indicate (and hence, it will not be very powerful).

Turning to the alternative of restrictive assumptions, the univariate approach is valid if the variables corresponding to the different levels of the repeated factor are uncorrelated and have equal variances. For the raw data this is clearly unlikely, but it might be true for transformed variables. In particular, note that if the raw variables have equal variances and if all pairwise covariances are equal (a condition known as compound symmetry) then the set of *deviation variables* – the difference between each variable and the average of all of them – satisfies this assumption. The normalized orthogonal polynomial components also satisfy the assumption. Indeed, any normalized orthogonal transformation will lead to new variables which satisfy the initial requirement. Transformations such as these are referred to as *orthonormal transformations*. (In fact, the condition of compound symmetry itself can be slightly relaxed, but we shall not pursue this point.)

In such circumstances, when the transformed variables are uncorrelated and have equal variances, the univariate approach is not only valid, it is also more powerful than the multivariate approach. Unfortunately, however, these conditions for justifying the univariate approach seldom seem likely to be true in the behavioural sciences (much less often than the univariate approach is in fact used!).

In general, it seems wiser and safer to adopt the multivariate approach. The reason that some textbooks (predominantly the earlier ones and the more elementary ones) recommend the univariate method is for its ease of calculation. The univariate method can be carried out on a hand calculator, whereas this is not feasible in the multivariate case. On the assumption that the reader has access to a multivariate analysis of variance computer package, there seems little justification for adopting the univariate method as a general strategy. We note in passing, however, that some computer programs do print out the significance tests appropriate to a univariate analysis, under the title 'averaged F-tests' – this being a reference to the computation of the sum of squares under the restrictive assumptions given above.

3.7 FACTORIAL STRUCTURES

Having established the principles of transforming the response variables, it is a simple matter to extend the discussion to response variables which are arranged in a more complex factorial structure. Just as in the between-subjects case, where a classification factor partitions the subjects into distinct groups or categories, so may a classification factor between variables (i.e. within subjects) partition the variables into distinct groups or categories. Thus in a repeated-measures study involving measurements at five time points we can regard time as a five-level, within-subjects factor. Furthermore, as with the between-subjects design, so the within-subjects design may have a cross-classified structure. There might be three types of measurement taken at each of the five time points – a type-by-time (three-by-five) within-subjects design. Any number (within the bounds of practicability) of factors may be involved in the within-subjects design. Note that 'time' is a commonly encountered factor, but it is not always present. For example, in a drug trial each subject might be given three doses of each of two drugs and have five different aspects of response measured: a three-by-two-by-five factorial within-subjects design.

When cross-classified factors are used to categorize the variables it is possible to think in terms of main effects and interactions of these within-subjects factors, just as earlier (Chapter 2) we thought of main effects and interactions of the factors which were used to group subjects. In fact, these within-subjects effects – both factor main effects and the interactions – will be represented by sets of variables which are derived from the original variables via a transformation using contrasts and combinations of the kind described in Chapter 2. There will be a set for the main effect of factor 1, a set for that of factor 2, and a set for the interaction of these two factors. If there are more than two factors, there will be sets of variables corresponding to main effects for each, first-order interactions, second-order interactions, and so on.

Now, each of these sets of derived variables can be used to analyse the between-groups comparisons. That is to say, for all comparisons that are made between groups, there will be a possible analysis of them using the variable set representing factor 1 main effects; there will also be another analysis of them which uses the set for factor 2 main effects, and another the set for the inter-actions, etc. Each of these will be analysing the same between-group comparisons using a different aspect of the information provided by the original variables, transformed through their factorial structure. An example is now given of a transformation which results from a two-factor within-subjects design on the response variables.

■EXAMPLE 1   To illustrate, we shall consider a fictitious study of a new dietary regimen. The primary aim will be to explore self-image and we will have two measures of self-image (perhaps scores on two different questionnaires). Since the change of attitude with time is of primary interest we shall take these two measurements at each of five time points during a year. The total of ten observations on each subject then constitutes a factorial design involving the crossing of two within-subjects (between variables) factors: *type* of measure-ment (two levels) by *time* measurements taken (five levels). To keep things simple we shall suppose we have just a single group of subjects (the ultimate simple between-subjects design).

As with the simple time trend example of Section 3.5, the actual observed variables are not of primary interest here. What are of interest are certain relationships between the measured variables. We shall therefore apply trans-formations to derive new variables embodying these relationships. In this

58

example, an initial question might be 'Do the two self-image measures evolve in the same way over time?'

Each of the two sets of five self-image scores can be converted into five alternative (polynomial component) measures, as before, which summarize more lucidly the patterns of change over time. Then a further transformation can be made in which, for each of the polynomial components, differences between the two self-image scores are calculated. Thus, for example, the difference between the first polynomial components of the two measures tells us whether the two scores have different overall levels (the first components being the overall averages for each type of score). The difference between the second polynomial components tells us whether the two scores have different linear trends (slopes): does one measure type change more rapidly than the other? The next difference compares rate of change of slopes, and so on.

The reader will probably have noticed that we began with ten scores (two self-image measure types at five time points) and yet have now only five scores (five polynomial component difference measures). If the transformation is to result in a set of measures equivalent to the original data we must have ten derived variables. The other five are obtained by summing (or averaging) each pair of polynomial components, instead of finding the differences. These sums measure overall self-image evolution over time (and would make limited sense unless no difference between the time behaviour patterns of the two self-image scores had been observed from the polynomial component difference calculations).

We can express this in terms of main effects and interactions as shown in Table 3.1. The ten original variables are labelled a to j and the ten derived variables A to J. The main effect of measurement type is simply the difference between the overall averages of the two types (F). It is thus expressed by a single variable. The main effect of time is the average pattern of change over time – new variables B, C, D and E summarize this. Variables G, H, I and J show how change over time differs between types – and are thus the interaction variables. Finally, A is the overall (time and type) average. This variable might be of interest in a multiple group study in which no time, type or interaction effects between variables were found. In such a case one might well wish to compare the overall levels of the groups, and variable A would be the appropriate variable to use.

The structure of the transformation from old to new variables is exactly the contrast structure which would be used for an analogous between-groups design. The complete comparison covering the whole cross-classification is the set of nine contrasts used to generate variables B to J. The comparison for the main effect of time is the set of contrasts for variables B to E, and so on.

Now that we have partitioned the variables into four blocks, A, B to E, F, G to J, we can conduct four separate analyses. Each analysis will answer a question about pattern of change over time, over type, or the interaction of such changes

Within-subjects analysis

(we are here ignoring *A* because, in this study, it is of limited interest). For sets *B* to *E* and *G* to *J* we might conduct intrinsically multivariate analyses – just to answer the questions 'are the average scores of the variables in these sets zero?'. Or we might wish to try to characterize any discerned pattern of change over time as linear, quadratic, etc. (Chapter 5 shows how to do this.) ■

In Section 2.15 we presented some common contrast sets which could be applied to each factor in a multi-factor design. In within-subjects designs polynomial contrasts are particularly common because of their role in repeated measures

Table 3.1 The time by self-image set of variables transformed into a more useful derived set.

	Time				
	1	*2*	*3*	*4*	*5*
Self-image type 1	*a*	*b*	*c*	*d*	*e*
Self-image type 2	*f*	*g*	*h*	*i*	*j*

	Polynomial component				
	Mean	*Linear*	*Quadradic*	*Cubic*	*Quartic*
Average of types 1 & 2	*A*	*B*	*C*	*D*	*E*
Difference of types 1 & 2	*F*	*G*	*H*	*I*	*J*

studies. Another common set involves comparisons between one level and all the others. The chosen one might, for example, be a baseline against which others are to be compared. We note, in passing, that whether or not a common baseline is subtracted has no effect on the linear, quadratic, etc., components of time change. This is because these components are defined in terms of contrasts whose elements sum to zero. The baseline score will be added and subtracted in such a way that its final weight is zero.

We should also remark that, as for the between-subject case, most computer programs for multivariate analysis of variance allow standard contrast sets to be chosen using some kind of simple naming or symbolic notation. It is not necessary for the researcher to specify the actual numeric elements, although there is usually provision to do so if desired.

Display 12 A factorial within-subjects design

In Display 1 we described a two-by-two within-subjects design involving the factors time pressure and mode of presentation of the problems (verbal or geometric). The four resulting variables, M1 to M4, are defined in Display 1. For simplicity we shall just consider the analysis of groups G1 and G2.

The research objectives might be summarized by saying that the experimenter wishes to see how time pressure, mode of presentation, and the distinction between lectures and practice affect the response. The first two of these factors are within subjects, while the third is between subjects. This is probably an appropriate

Table X.12.1 Within-subjects factorial data for Display 12. (See Display 1 for definitions of variables.)

	M1	*M2*	*M3*	*M4*
G1	10	14	13	12
	9	15	10	12
	7	13	10	16
	11	13	9	15
	13	20	8	15
G1	16	16	13	11
	16	17	13	12
	13	16	9	16
	7	19	17	25

Table X.12.2 Anova results for within-subjects factorial example.

	Sum of squares	d.f.	Mean square	F	P
Constant	6426.694	1	5626.694	977.975	.000
Condition	55.556	1	55.556	8.454	.023
Within	46.000	7	8.454		
Presentation mode	2.250	1	2.250	.118	.741
Present × Cond	.000	1	.000	.000	1.000
Within	133.000	7	19.000		
Time pressure	148.028	1	148.028	10.907	.013
Time press × Cond	2.222	1	2.222	.164	.698
Within	95.000	7	13.571		
Prest × Time press	2.250	1	2.250	.716	.425
Prest × Time press × Cond	.000	1	.000	.000	1.000
Within	22.000	7	3.145		

moment to remind the reader that the Displays in this text do not in any way strive to present real problems. Thus in this particular example we will not discuss order and learning effects, etc. The data for this example appear in Table X.12.1.

The following contrasts applied to M1 to M4 yield, respectively, derived variables showing average response, the effect of mode of presentation, the effect of time pressure, and the presentation by time pressure interaction (or scaled version of these):

1	1	1	1
1	1	-1	-1
1	-1	1	-1
1	-1	-1	1

The results are shown in Table X.12.2. This table falls into four blocks, each one corresponding to one of the four derived variables in the order given above.

To interpret it we start, as always, with the highest-order effect, working our way back to lower-order ones until a significant effect is found. In this case the three-way interaction is not significant, and neither are any of the three two-way interactions. However, two of the main effects, those for Condition and Time pressure, are significant at the 5% level.

3.8 CONTROLLING FOR OTHER VARIABLES

Readers familiar with univariate analysis of variance may also have encountered *analysis of covariance*. This is analysis of variance in which differences between groups are tested controlling for other variables, termed covariates. That is, we use the (estimated) relationship between the response variable and the covariates to adjust the subjects' response to the scores they 'would have had', had all covariate scores been the same; and only then test to see whether the groups differ. Note that introducing a covariate into an analysis might remove apparent between-groups differences or it might reveal differences which were previously concealed. These are both instances of the more general idea that differences between groups of individuals can be distorted by an inequitable distribution within the groups. The principle was illustrated in Section 2.8 in relation to treatment groups and sex.

In either case we could express the purpose of covariance analysis as being to see if the differences between subjects, as measured on the covariates, were adequate to account for the observed differences between their responses, or whether we must conclude that the response variable is showing differences over and above those attributable to the different covariate scores. Analysis of covariance can also increase the power of tests by reducing the within-group variation – in exactly the same way as can introducing extra classifying factors (see the end of Section 2.10).

Introducing covariates into a multivariate analysis of variance is, in principle, no more complicated than using them in univariate analysis. Again the question being asked is whether or not the covariates are adequate to explain the observed response differences between (groups of) individuals. The major distinction, of course, is that the between-groups differences are now defined in terms of vectors of means rather than single means. However, covariance analysis does play a more prominent role in multivariate analysis because of the prevalence of questions involving sequences of variables – that is, of variables or sets of variables which fall into a natural order. Typically one of the variables, the last in the sequence, is chosen as a response variable and the between-groups effects are calculated for this variable after 'eliminating' effects due to the others through analysis of covariance. If no effect is discovered, this response variable is dropped from the analysis and the process is repeated using the next in the sequence.

Since such sequences of variables arise from many of the transformations we have been discussing, for example polynomial components, or variable sets in within-subjects factorial designs, a natural question is whether or not to use this covariance approach here. That is, if a within-subjects design has been transformed to yield polynomial components of time change, should the mean and linear components be 'covaried out' when testing the quadratic, should the mean be 'covaried out' when testing the linear?

The two approaches, covarying and not covarying, in fact address different questions, and the choice must lie with the researcher and his or her aims. In our experience the behavioural sciences more usually do not require covariance analyses of this sort. In the case of polynomial change components, for example, we will often want simply to address the question 'do the groups have differing average slopes, or are they the same?', and this implies testing the linear component without using the mean level as a covariate. If we do use it as a covariate, we are asking the question 'can the differences in slope be predicted from the differences in mean level?' (i.e. 'if we adjust all subjects to the same overall mean score, will the groups *then* have the same average slopes?'). The ability to predict the slopes in this way does not tell us that such differences do not exist; but it may indicate that a simpler parameterization of the data could be constructed. Of course, there may be exceptional cases where this second question is the more pertinent one of the two. The same principle applies to whether or not to use as covariates the main effects variables (for example) when testing interaction variables in a within-subjects factorial design. We shall discuss this issue further in Chapter 5.

Display 13 A univariate analysis of covariance

One researcher was uneasy about two aspects of the data. Firstly, the large amount of intrinsic variation between subjects might swamp real differences between condition groups unless excessively large samples were taken. It was therefore necessary to increase the power of the tests. One way this could be done was by taking account of subjects' basic ability and somehow controlling for it.

63

Within-subjects analysis

The second concern was that the groups might differ before the experiment. Although the subjects had been randomly assigned to groups, there were only small samples and it was possible that their initial means might differ. Again this might be overcome by controlling for the basic ability.

To accomplish these ends problem-solving skill was measured prior to any exposure or practice, producing a variable B (for baseline), and it was used as a covariate in the analysis of M1. Again, interest lay in the contrasts

$$\begin{matrix} 1 & 0 & -1 \\ 0 & 1 & -1 \end{matrix}$$

between the three groups, used in a complete comparison.

The raw data are shown in Table X.13.1. Table X.13.2 shows an anova table without the covariate. Contrast 1 appears highly significant, while Contrast 2 does not appear at all significant.

Now what happens when we introduce the covariate? This is shown in Table

Table X.13.1 Raw data for the analysis of covariance in Display 13.

	B	M1
Group 1 (Lectures)	11	9
	13	12
	11	7
	10	7
	13	10
Group 2 (Practice)	9	12
	13	15
	10	15
	14	18
Group 3 (Control)	17	18
	16	14
	14	14
	15	14

Table X.13.2 Anova table for Display 13.

	Sum of squares	d.f.	Mean square	F	P
Constant	2172.857	1	2172.857	452.679	.000
Contrast 1	80.000	1	80.000	16.667	.002
Contrast 2	.000	1	.000	.000	1.000
Within	48.000	10	4.800		

Table X.13.3 Effect of introducing baseline as a covariate.

	Sum of squares	d.f.	Mean square	F	P
Regression	32.911	1	32.911	19.630	.002
Constant	.076	1	.076	.045	.836
Contrast 1	3.562	1	3.562	2.124	1.79
Contrast 2	17.208	1	17.208	10.264	.011
Within	15.089	9	1.677		

X.13.3. We notice first that B and M1 have a highly significant relationship. We also notice a complete turnaround of the results on the two contrasts. Groups 2 and 3 had identical M1 means prior to adjusting to equal B means, and this has produced a significant M1 difference. In contrast, the difference between groups 1 and 3, previously highly significant, now becomes non-significant.

A further interesting result here is that the constant term, previously highly significant, becomes completely insignificant when the covariate is introduced. This is because when estimating the constant term using the covariate, all groups are adjusted to a zero covariate score. (Alternatively, one can think of the constant as the intercept on the M1 axes, when B = 0.) Care should be used when interpreting the constant term in an analysis of covariance since different packages may make different adjustments. For example, whereas this package adjusts to B = 0, others adjust to the grand mean of B.

3.9 INTRINSICALLY MULTIVARIATE STUDIES: TWO GROUPS

This is the third type of multivariate question identified in Section 3.2. Here we are simply asking whether there are any between-groups differences, regardless of which individual variables may be responsible for the differences. In some ways such questions are simpler than the other kinds: they do not involve specifying appropriate combinations of variables or deciding whether to estimate the effects of some while controlling for other variables. But in other ways they are more complicated: the questions are more fundamentally multivariate, so that new ideas must be brought to bear and there is greater scope for choice.

We shall present the basic ideas by way of a simple example involving just two groups measured on four variables. The question we wish to answer is simply 'do the groups differ?' with no consideration of how they differ, if they do. (Such considerations may arise later, and are the province of Chapter 5.)

From Chapter 2 we know that an appropriate between-groups contrast for answering this question is $(+1 \ -1)$. That is, we apply element $+1$ to one group and -1 to the other. Now, what exactly are the measurements to which we should apply these contrast elements? If the problem had been univariate, involving a single response variable, we would simply apply the contrast weights to the group means, producing a straightforward difference between means which could be examined by a t-test (or, squaring it, an F-test). In the multivariate case an analogous application of the contrast to the vector of group means yields a vector of four differences. How are we to obtain from these four difference measures an answer to our question?

We could, of course, conduct four separate univariate analyses in the usual manner, and if a significant difference were found on any one of the variables then we might conclude that a difference between the groups existed. This seems simple enough, so we must ask why it is necessary to do anything more complicated. What is wrong with this straightforward multiple univariate approach? What advantages are there to any alternatives?

There are two main advantages in adopting an approach which is more fundamentally multivariate than the four separate univariate analyses. The first is, as we have already mentioned in Section 3.3, the fact that when we conduct several tests there is an increased chance of incorrectly rejecting a null hypothesis. The more tests one does, the greater this chance becomes so that we become more and more likely to generate a significant result by chance alone. We discuss in detail in Chapter 5 how we can use multivariate techniques to control this problem.

A second advantage, and one which will be the focus of the remainder of this chapter, is gained from being able to use the information in the data more completely. It might be the case that a combination of variables is necessary to show properly the difference between the groups. Figure 3.1 illustrates this for the case of two variables. Both variables 1 and 2, when taken individually, as shown by the marginal plots, show considerable overlap between the groups. If we looked at either of the variables individually our samples might not show significant separation between the groups. And yet the groups clearly are separated, as the bivariate plot shows. What we are after, then, is some way to tap the clear separation which does exist in the multivariate (here bivariate) space. In the bivariate case we can plot scattergrams of the two groups, but what shall we do in cases involving more than two variables?

Figure 3.1 gives us a hint. Let us examine the distribution of the new variable defined as the *sum* of the original two. The distributions of this variable for the two groups are shown along the line AA′ in Fig. 3.1. It is apparent that for this new, derived variable there is no overlap at all between the two groups. That is, variable 1 plus variable 2 produces a new score for which members of group 1 always have lower values than members of group 2.

This, then, is the general principle: we try to identify a derived variable, in general a *weighted sum* of the original variables, on which the groups are well separated. Allowing a general weighted sum to be used, and not just the simple, unweighted sum used as an illustration in Fig. 3.1, means that we can try to identify a set of weights to *maximize the distance* between the two groups on the resulting derived variable. 'Distance' here will mean the difference between the

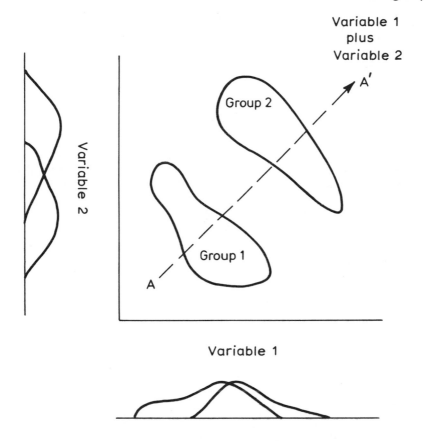

Figure 3.1 A clear difference between the groups can be discerned from the bivariate contours, but neither of the two variables alone shows such.

means of the groups on the new variable relative to the within-group standard error of the new variable. We could, of course, equally well maximize the squared value of such a distance – the same set of optimal weights for the weighted sum would result. This 'squared distance' approach thus finds the set of weights which leads to a variable on which the squared difference between the means divided by the variance within groups is maximum. That is, in terms of the discussion in Chapter 2, it maximizes the ratio of an attributable squared term to an error sum of squares.

The only difference between this multivariate situation and the univariate analysis of Chapter 2 is that we are now seeking a weighted sum to maximize the measure of difference between groups. As a consequence of this, the testing procedure (to be discussed in Chapter 4) will be slightly different from the simple procedures of Chapter 2.

Display 14 An intrinsically multivariate test involving two groups

---

So far we have not focused attention on the three creativity measures described in Display 1. Let us do so now. These measures are not of any great intrinsic interest in their own right: it is as a whole, as a set of three, that they matter. Thus our researcher does not wish to examine each of them: rather, to perform some global multivariate test of between-groups differences. We will suppose that a comparison is to be made between groups G1 and G2 (lectures and practice). The data are given in Table X.14.1.

Simple univariate tests on C1, C2 and C3 give P-values 0.587, 0.262 and 0.323 respectively. From this one might conclude that the groups did not differ. A global multivariate test, however, gives $P = 0.074$. Depending on one's choice of significance level, this might be regarded as significant.

This example continues in Display 16.

Table X.14.1 Multivariate creativity data for Display 14.

Group 1				Group 2		
C1	*C2*	*C3*		*C1*	*C2*	*C3*
7	15	12		6	12	14
9	13	10		10	10	14
10	8	11		20	9	16
10	8	14		8	17	8
14	6	13		8	15	13
8	14	11		14	9	13
9	12	11				
10	10	12				
11	8	13				
12	6	13				

3.10 INTRINSICALLY MULTIVARIATE STUDIES: SEVERAL GROUPS

Given the parallel between the multivariate and univariate cases outlined at the end of the preceding section, the way to proceed for any other single contrast is clear. We simply try different sets of weights to find that set which maximizes the ratio of an attributable squared term to an error sum of squares.

We can extend this idea to more-general comparisons of several groups, involving more than a single contrast. When we examined this in the univariate case we obtained test statistics involving an attributable *sum* of squares for the complete comparison and we calculated the ratio of this attributable sum of squares to an error sum of squares. Now we seek to find a single new variable, derived as weighted sums of the original variables, for which this same ratio of attributable sum of squares to error sum of squares is a maximum. In general, for an arbitrary comparison, the optimum weighted combination is called the *first canonical variable* for that comparison.

This name suggests that there might be other canonical variables for the comparison, and indeed there are. We can define a second canonical variable in the same way as the first: it is that weighted sum of the variables which maximizes the ratio of the attributable sum of squares to error sum of squares *subject to the restriction that it is uncorrelated with the first*. This can, of course, be repeated, at each step requiring that the new variable is uncorrelated with those already chosen.

There is, however, a limit to how far we can go like this. Firstly, we cannot extract more canonical variables than there are original variables. Secondly, we cannot extract more canonical variables than there are degrees of freedom in the comparison. For a single contrast, for example (as in Section 3.9), we can only extract a single canonical variable. These two limitations together tell us that the maximum number of canonical variables that we can extract is equal to the minimum of the number of original variables and the number of degrees of freedom in the comparison.

There are several interesting things to note about these canonical variables. First, if we take them all, then they provide exactly the same information as was given by the original variables. That is, if we applied the chosen weights to the original variables to compute scores on the new variables for each subject, these scores would be an equivalent set of data to the original set. As a sort of sequel to this, if we only took the first few, say, the first two canonical variables, then this would be the best possible set of two (in this case) variables with which to summarize the between-groups difference pattern described by the comparison. No other two variables would be as good. This generalizes to the first three, the first four, etc.

Note also that the canonical variables are derived by applying weights to the original variables – and that is just what we did in the structured multivariate case of Sections 3.4–3.7. The difference is that now we choose what weights to apply by calculations made from the data themselves. One consequence of this is that tests of significance will be different in the two cases. Another consequence is that canonical variables can be difficult to interpret. They do not have the simple contrast weights, chosen by the researcher, of the structured case.

We should also mention two other terms, since they appear in computer output. The *first canonical correlation* is the square root of the ratio of attributable to total sum of squares (where total sum of squares is the sum of the attributable and within-group sums of squares) for the first canonical variable. This can be extended to the second canonical correlation, and so on. Similarly, the *first eigenvalue* is the ratio of the attributable sum of squares to the within-groups sum of squares for the first canonical variable, and this can also be extended to subsequent eigenvalues.

Within-subjects analysis

An intrinsically multivariate test on a comparison

Here, as in Display 14, we again consider the three creativity measures on a unified set. Now, however, instead of looking at a particular contrast, we are going to examine a set of contrasts – a comparison. In fact, we shall look at the complete comparison on the three groups G1, G2 and G3. The data are given in Table X.15.1.

The new derived variables – the canonical variates, which best separate the groups in the way described in the test – are given by applying the weights in Table X.15.2 to the three creativity measures. The eigenvalues and canonical correlations are given in Table X.15.3.

This example continues in Display 16.

Table X.15.1 Multivariate creativity data for Display 15.

	C1	C2	C3
Group 1	10	15	7
	9	15	7
	11	10	7
Group 2	15	13	17
	13	22	18
	15	17	17
Group 3	15	13	17
	17	18	16
	19	20	18

Table X.15.2 Weights for the canonical variates.

	C1	C2	C3
Canonical variate 1	.021	−.167	1.654
Canonical variate 2	.693	−.062	−.327

Table X.15.3 Eigenvalues and canonical correlations for the data from Table X.15.1.

	Eigenvalue	Canonical correlation
Canonical variate 1	87.573	.994
Canonical variate 2	1.752	.798

3.11 FURTHER READING

Univariate repeated measures analyses are well described in Winer (1971) and Keppel (1973). More-mathematical descriptions of between-variables structures are given in some of the textbooks described in Chapter 5, notably Finn (1974), Bock (1975) and Morrison (1967). For between-variables analysis on commensurable variables (i.e. those measured on the same scales), which are often, but need not be, repeated measures situations, some books use the term *profile analysis*.

Multivariate ————————————— 4
tests

4.1 INTRODUCTION

In Chapters 1 and 3 we discussed reasons for adopting multivariate techniques. We found it useful to classify these questions into three categories:

(a) Multiple univariate studies, in which the researcher is interested in a number of separate and clearly identified response variables.
(b) Structured multivariate studies, in which concern is with new variables derived as weighted sums of the original raw variables.
(c) Intrinsically multivariate studies, where concern is with comparisons measured on sets of variables simultaneously, and where one is not interested in predefined contrasts on the individual variables.

As we pointed out in Chapter 3, type (a) questions are relatively straightforward. They reduce to a number of distinct univariate analyses. The only complication arises regarding an appropriate choice of significance level – the type I error rate – for the analyses. Should each be conducted separately at the 5% level, say, or should an overall protection (against *any* error of type I) at the 5% level be used? The answer depends on the types and sizes of risks that the experimenter is prepared to take, but, given that separate and distinct questions are involved, a case can certainly be argued for adopting a 5% level for each variable.

Type (b) questions, after reformulation in terms of the derived variables, become either type (a) or (c) (or both) so that nothing intrinsically new arises here.

Type (c) questions are often followed by sequels: having established that some comparison is significant on a set of variables, then one may ask which variables account for the significance and on which contrasts they are significant. These questions are different from types (a) and (b) because they (at least implicitly) involve a search through all possible sets of variables and all possible contrasts. Such questions are answered using simultaneous test procedures, which are outlined in Chapter 5. This chapter deals with type (c) questions themselves: is a between-groups comparison significant when measured on several response variables simultaneously?

4.2 TOWARDS MULTIVARIATE TESTS

In Section 3.10 we described how the first canonical variate for a comparison is a new variable, derived as that weighted sum of the raw variables which maximizes a particular ratio: the ratio of its attributable sum of squares to its error sum of squares. The second canonical variate was defined in an identical way, with the extra qualification that it was uncorrelated with the first; and so on. Associated with these canonical variates will be numbers indicating the actual sizes of these ratios of sums of squares.

In the univariate case all this simplifies: there is a single canonical variate, and that is the single variate itself. The attributable and error sums of squares are calculated in a straightforward manner (no process of maximization is needed, of course). In Section 2.6 we reminded the reader how the univariate attributable and error sums of squares are combined, by calculating their ratio, to yield a statistic which could be used to test the null hypothesis of a zero comparison. In the multivariate case we shall perform an analogous exercise, but complications arise since we now have several such ratios, one for each canonical variable. We are thus confronted with the problem of how to combine these separate ratios into a single test statistic in an effective way. This will be the topic of the next section. Before we turn to that, however, some further discussion of the sums of squares ratios is in order.

A property which is obviously desirable in a test statistic is that it should be invariant to rescaling transformations. This simply means that we would like to get the same result whatever the units in which our measurements were taken. A test which yielded different significance levels according to whether height was measured in centimetres or inches would be of somewhat limited value. In the univariate case the usual F-test is the uniformly most powerful test which is invariant to rescaling transformations (assuming, of course, that the data meet the usual distributional requirements for a valid F-test). In the present multivariate case, where we want to examine a group of variables simultaneously, it is necessary to extend the invariance principle to transformations which include not only any rescaling but also any re-expression in terms of weighted sums of the raw variables. (For example, we would hope to get the same multivariate test result whether the data were presented in terms of measurements at times 1 and 2 or in terms of a measurement at time 1 and a change to time 2.) Unfortunately, in the multivariate case there is no one uniformly most powerful test which is invariant to all these transformations. Which test is best, in the sense of being most powerful, depends on the nature of the departure from the null hypothesis. This lack of a uniformly most powerful test is one of the reasons why several different test statistics have been proposed.

However, even if the invariance requirement has not narrowed the choice down to one test statistic, it does help. It can be shown that a test statistic based on the ratios of the sums of squares for the canonical variates, as outlined above, is invariant to rescaling and re-expression in terms of weighted sums. Thus use of these ratios is clearly a natural way to proceed.

An alternative way of approaching these ideas is as follows. In the univariate case the test statistic is based on a ratio of H, the attributable sum of squares, to E, the error sum of squares; that is, on (H/E). In Section 3.1 we described how in the multivariate case H generalizes to become an attributable sum of products matrix. A similar generalization applies to the error sum of squares, which becomes an error sum of products matrix in the multivariate case. If we now let $\mathbf{H}$ and $\mathbf{E}$ represent these matrices, the matrix equivalent of the above ratio is $(\mathbf{H \cdot E}^{-1})$. The sum of squares ratios for the canonical variates, or the *eigenvalues*, as we termed them at the end of Section 3.10, can then be directly calculated from this matrix.

4.3 MULTIVARIATE TEST STATISTICS

Many ways of combining the eigenvalues (the ratios of sums of squares corresponding to the canonical variates) have been considered. We shall merely consider the four which appear most frequently in computer packages. In decreasing order of sizes let the eigenvalues be denoted by $\lambda_1, \lambda_2, \ldots$ (so that λ_1 is the sum of squares ratio for the first canonical variate). Then the four test statistics are:

(1) The Pillai–Bartlett trace: $\Sigma \lambda_j / (1 + \lambda_j)$.
(2) Wilks's lambda: $\Pi (1 + \lambda_j)^{-1}$.
(3) The Hotelling–Lawley trace: $\Sigma \lambda_j$.
(4) Roy's largest eigenvalue statistic: λ_1.

When the number of variables is only one – the univariate case – then all four criteria yield equivalent results, being simply transformations of the usual F-statistic. Also, when the hypothesis degrees of freedom are exactly one, as in a simple two-group comparison, all four statistics yield equivalent results.

In the general case the distributions of statistics (1) to (4) are complicated: they are not simple F-distributions or anything like that. However, by applying appropriate mathematical transformations, statistics can be derived which approximately follow chi-squared or F-distributions. It is these transformed statistics which yield the test results printed in computer output listings.

Those who already have some knowledge of multivariate statistics might be interested to note that criterion (2) is the likelihood ratio statistic for testing that the hypothesis comparison is zero, assuming multivariate normality and further assuming equality of the covariance matrices for all groups. The appellation 'trace' for criterion (1) may puzzle those readers familiar with matrix algebra, since it is not the sum of the eigenvalues of the matrix $\mathbf{H \cdot E}^{-1}$. In fact it is the trace of the matrix $\mathbf{H \cdot (H + E)}^{-1}$ which is the matrix analogue of the ratio of the hypothesis sum of squares to the total sum of squares. This, of course, contains the same information as the matrix $H \cdot E^{-1}$. Criterion (4) is often written, too, not as the largest eigenvalue of $H \cdot E^{-1}$ but instead as the largest eigenvalue of $H \cdot (H + E)^{-1}$; it is known as Roy's maximum characteristic root or as the union–intersection statistic.

Apart from the question of which of these statistics to use, the discussion above basically provides the procedures we need for the overall testing of intrinsically multivariate questions. To discover if some contrast or set of contrasts differ significantly from a particular value or values we (or rather, a computer program) simply combine the values λ_1, λ_2, etc., appropriately, using one of methods (1) to (4), transform the result so that it can be referred to an F-distribution, and compare this transformed result with the appropriate table. We discuss choice of criterion in Section 4.4.

Display 16 Multivariate test statistics

Some computer packages (e.g. the SPSS-X MANOVA) give the user a choice of multivariate test statistics. Others simply present the results of an analysis using one statistic.

First consider the data given in Table X.14.1. A test on the difference between those two groups using all three creativity variables leads to Table X.16.1. We see that the four test statistics are identical.

Now consider a three-group example using the data of Table X.15.1. The complete comparison on three groups has two degrees of freedom, and we see that in such a case (Table X.16.2) the four test statistics do not yield identical results.

Table X.16.1 Multivariate test statistics and significance levels for the data of Table X.14.1.

Test name	Value	Approx. F	Hypoth. d.f.	Error d.f.	P
Pillai	.42629	2.972	3.00	12.00	.074
Hotelling	.74304	2.972	3.00	12.00	.074
Wilks	.57371	2.972	3.00	12.00	.074
Roy	.42629				

Table X.16.2 Multivariate test statistics and significance levels for the data of Table X.15.1.

Test name	Value	Approx. F	Hypoth. d.f.	Error d.f.	P
Pillai	1.62534	7.23	6.00	10.00	.003
Hotelling	89.32536	44.66	6.00	6.00	.000
Wilks	.00410	19.48	6.00	8.00	.000
Roy	.98871				

4.4 CHOICE OF TEST STATISTIC

One of the most important considerations in selecting a test statistic is the power that it has to detect that the alternative hypothesis is true. Unfortunately, the relative power of the four criteria introduced in Section 4.3 depends on the way in which the group mean vectors in the underlying populations depart from the null hypothesis of equality. The mean vectors might show a consistent trend for each variable so that, approximately, they all constitute only a single dimension. Such an arrangement is termed a *concentrated noncentrality structure*. In contrast, there might be departure from equality in all directions, so that relationships between the mean vectors of the groups exhibit no clear structure. This is a *diffuse noncentrality* structure.

Chester L. Olson has made a comparative study of the four statistics introduced in Section 4.3. He found that for concentrated noncentrality structures the power decreased in the order (4), (3), (2), (1), while for diffuse structures the order was reversed, with (1) being the most powerful. Since it will be a rare study in which the investigators will be able to specify precisely what kind of departure from the null hypothesis is to be expected under the alternative, we are presented with something of a dilemma. However, it seems that the concentration into one dimension must be fairly extreme before it changes the power ordering. In view of this Olson favours statistic (1).

Power is just one basis for choosing between the criteria. As in the univariate case, the test statistic distributions will be based on certain assumptions about the basic data distributions, namely independent multivariate normally distributed observations with identical covariance matrices in each group. Insofar as the data do not conform to these assumptions, we might expect to obtain invalid results. Thus, robustness to violation of these assumptions is also an important comparative basis for choice. In his studies, Olson found that occasional extreme observations have mild effects on type I error rates, tending to make the tests more conservative rather than less. However, of the four criteria above, (1) seems to remain closest to the nominal significance level. Insofar as hetero-scedasticity (inequality of variance) is present, Olson found that departure from homogeneity of the covariance matrices can have severe effects on (2), (3) and (4), but a much less severe effect on (1). In all four cases, however, such departure led to increased type I error rates.

We shall have more to say about the relative merits of these various criteria in Section 5.7, after we have discussed simultaneous test procedures. In that section we also make some recommendations.

Perhaps we should acknowledge that for many researchers the choice may be taken out of their hands by the computer package they are using. In this case it is still just as important to state, in any description of the study, which statistic was used.

4.5 EXPLORING NONCENTRALITY STRUCTURES

Since different test criteria have different power under different types of departure from the null hypothesis of equal group mean vectors, it is of general interest to know, when the null hypothesis is rejected, what kind of departure from the null has caused this. For example, is the noncentrality structure highly concentrated or not? In particular research studies this question may have special relevance – for instance, when the groups represent different stages or levels of some process, the variables might measure a single changing phenomenon, or might measure two differently developing aspects of the process.

The question can be tackled by testing to see how many of the canonical variables are needed to distinguish the groups: in the concentrated noncentrality structure only one variable is necessary. The question is equivalent to testing the size of the eigenvalues of $H \cdot E^{-1}$ to see how many are important. To see whether all beyond some specified number can be ignored, we can use Bartlett's test. This simply examines that part of criterion (2) (the likelihood ratio statistic) which is due to the canonical variates whose significance we wish to assess. Often a sequence of such tests is carried out and is referred to as a *dimension-reduction analysis*. Initially a test that none of the canonical variates is important for group separation is made. If this is rejected then a test is made to see if one is adequate. If it is not, then two are examined, and so on. Note that, as described previously, when a sequence of tests is carried out in this way we must consider the overall significance level as well as the significance level of each individual test.

Display 17 Exploring noncentrality structures

Tables X.17.1 and X.17.2 give two data sets, each showing two variables measured on the subjects in four groups. The eigenvalues from Table X.17.1 are 4.391 and 0.006, while those for Table X.17.2 are 7.372 and 4.631. It is clear that in the first case a single dimension is adequate to represent the data, while in the second two dimensions are needed.

The reader will gain insight into the mechanism of what is happening here if he or she plots the means of the four groups on a graph using the variables as axes.

Table X.17.1 Data for Display 17 exhibiting concentrated noncentrality structure.

G1		G2		G3		G4	
V1	V2	V1	V2	V1	V2	V1	V2
10	14	16	16	6	17	18	21
11	8	15	16	22	20	22	19
12	8	17	16	26	14		

Table X.17.2 Data for Display 17 exhibiting diffuse noncentrality structure.

G1		G2		G3		G4	
V1	V2	V1	V2	V1	V2	V1	V2
10	14	7	25	22	8	18	21
11	8	6	19	26	8	22	19
12	8	13	21	17	11		

4.6 COMPLICATIONS: THE HYPOTHESIS MATRIX

In Chapter 2 we described between-groups comparisons and the general problems of non-orthogonality and of controlling for the effects of some factors in the analysis. In multivariate analysis, the situation is exactly analogous to the univariate case. There, except in a balanced design, the simple hypothesis sum of squares computed separately for each factor, or indeed for any comparison, is used to answer one sort of question; a different question is associated with the sum of squares computed after adjusting for the effects of other factors in the design. The only change we need to make in a multivariate analysis is that it is now an hypothesis sums of products matrix which is adjusted for these other effects.

Thus the same exception applies: in a carefully designed and balanced experiment, it may be that some between-group comparisons (these usually being the complete comparison for each factor's main effects and the interaction comparisons, or particular contrasts of special interest) will yield the same hypothesis sum of products matrix whether or not we control. However, in general in unbalanced designs, the questions we want to answer mean that the between-groups comparisons must be made controlling for other factors' effects. We must derive an hypothesis matrix (via a computer program, of course) which is suitably adjusted for these other effects before the test procedures described above can be applied. For example, in a design involving a cross-classification of the groups by two factors, comparisons between levels of one factor will probably be made controlling for the effects of the other.

Note that the hypothesis sum of products matrix and the error sum of products matrix are constructed using precisely the variables we are investigating. These may be the original variables or some specific transformation of them. When there is a between-variables design there will be subsets of the transformed variables which represent each main effect and subsets which represent interactions in this design, as described in Section 3.7. Each of these subsets may be used separately to construct hypothesis matrices. In some cases these variables will be adjusted for the effects of others through analysis of covariance as discussed in Section 3.8. In any case, in all situations we will use the transformed response variables to construct relevant hypothesis and error sums of products matrices.

4.7 COMPLICATIONS: THE ERROR MATRIX

In practice, most analyses use the within-group sum of products matrix as the 'error' matrix in calculating the ratios of hypothesis to error sum of squares. If covariates have been used then both matrices are based on adjusted variables, as described in Section 3.8. Sometimes, however, this is not the appropriate error matrix. The simplest example of a situation for which this is true is the univariate mixed two-way analysis of variance with, say, columns being a fixed effect and rows a random effect. In this case the test for a column effect is based on the ratio of column sum of squares to interaction sum of squares. In any case, even in a fixed effects model one might choose a different error matrix, combining the within-groups matrix with matrices arising from the assumptions that certain contrasts are zero, as described in Section 2.6. Most programs use the within-group sum of products matrix as the default matrix and permit the user to specify other choices if so desired. The theory is described in texts on univariate analysis of variance and generalizes immediately to the multivariate case.

A further situation in which the usual within-groups matrix would not be used arises when something is known about what the matrix ought to be. For instance, in some experiments designed with repeated measures, it might be reasonable to assume that the error matrix has only diagonal terms and that these are all equal. This can arise after appropriate transformation of repeated measures for which the covariance matrix of the raw variables shows 'compound symmetry' – the covariances between all raw variables are equal and so are all the variances. Then it is a more powerful procedure to use what in this context is referred to as an averaged F-test: the diagonals of the two matrices are summed and then tested against each other using their summed degrees of freedom. Such a method reproduces the univariate analysis of variance for repeated measure described in Section 3.6. Note, however, that the increase in power of this approach is bought at the expense of making further assumptions about the data; and as we pointed out in Section 3.6, we believe that the assumptions are rarely realistic.

4.8 TESTS ON MULTIVARIATE ANALYSIS OF VARIANCE ASSUMPTIONS

The primary objective of multivariate analysis of variance is to explore contrasts in the mean vectors. So far, all the tests we have described have been directed towards this end. However, there are other hypotheses which we may wish to investigate – most commonly those arising in relation to the basic underlying assumptions of the method.

One such assumption is that of equal within-group covariance matrices – directly analogous to the homogeneity of variance assumption in univariate analysis. Tests for this exist and some computer programs give them. However, it has been pointed out, at least for the univariate case, that such tests tend to be more sensitive to departure from normality than the basic test of the analysis of variance. This means that the more sensitive screening test may prevent one from carrying out an analysis which would have been relatively acceptable. When the assumption of equal covariance matrices across groups cannot be maintained, the analysis becomes a generalized Fisher–Behrens problem, which has a literature of its own.

4.9 TESTS ON MULTIVARIATE ANALYSIS OF COVARIANCE ASSUMPTIONS

An assumption made in analysis of covariance beyond those assumptions common to analysis of variance is that the relationships between the response ('dependent') variable and the covariates are the same in each group, apart from possible differences of the group means. A test for such an assumption is based on the following principle. We can define easily calculated numbers (derived from squared multiple correlation coefficients) to measure the unaccounted variation in the relationship between the response variable and the covariates and factors in two situations: assuming, first, that the relationship between the response variable and the covariates is the same in each group; and second, that the relationships may be different. The difference between these two numbers can then be used as a measure of the increased unaccounted variation, or error, which results from requiring identical relationships. If the increase in error is small compared with the size of the general unexplained variation (as derived from the case where the identity assumption is not made) then clearly the assumption of identical relationships is reasonable.

In the multivariate case we have a direct extension to matrices measuring the residual covariation between response variables after controlling for the covariates by the two different approaches.

4.10 FURTHER READING

The question of which test statistic to choose is a difficult one. Gabriel (1968) stresses the virtues of statistic (4); Olson (1974, 1976) presents simulation comparisons of several, finally preferring (1); Stevens (1979) challenges his conclusions and is countered by Olson (1979). Bird and Hadzi-Pavlovic (1983) stress that both overall and subsequent tests must be considered in the choice. We consider this matter further in Chapter 5.

Formulae relating to the test statistics described in this chapter may be found in the texts listed in Chapter 1.

Sequential and simultaneous test procedures ——————— 5

5.1 RESOLUTION OF GENERAL QUESTIONS

We began our discussion of the multivariate aspects of analysis of variance by delineating three types of questions that we might like to answer: (a) multiple univariate questions, (b) structured multivariate questions, (c) intrinsically multivariate questions. In the last case one is addressing the question of whether or not a comparison is significant on a group of variables simultaneously, and the individual identity of the variables is regarded as irrelevant to the basic question. However, should the intrinsically multivariate question show significance, then one might well be interested in exploring further, to ask the questions: on which variables or combinations of variables is the comparison significant, and which contrasts within the comparison are significant? In this chapter we turn our attention to such follow-up questions.

In univariate analysis of variance this is the area covered by multiple comparison techniques (such as those due to Tukey, Scheffé, Duncan and Aitkin) when, having found a significant overall F-test, we wish to explore which contrasts between groups are responsible for the difference. We can describe such techniques as permitting us to *resolve* the significant complete comparison into component contrasts: the overall null hypothesis (the complete comparison) stating that there are no between-group differences at all has been rejected and we wish to know which patterns of differences are non-zero. Sometimes we will be interested in all sub-comparisons and sometimes only in a restricted set of them – perhaps, for example, in all pairwise group differences. Sometimes, having established that some overall differences exist, we will test a nested sequence of comparisons so that we can derive a simplified description of the pattern of differences. An example of such a sequence is contrast set 5 in Section 2.15 – the polynomial contrasts, which describe a trend pattern across groups using models of increasing complexity.

5.2 ERROR INFLATION

Our main purpose in studying these techniques is to control the inflated chance of making an error when many tests are performed. Specifically, we are

81

concerned with the type I error rate, which is the chance of finding a significant difference when one does not, in reality, exist; i.e. of rejecting a null hypothesis that is in fact true. That the problem of multiple testing is more than just a straightforward one of combinatorial complexity can be seen from the following argument. Suppose two groups are involved, with 20 variables, and we have carried out a general test, as described in Chapter 4, and found an overall difference between the groups. We now wish to know which variables are responsible for the difference. For the sake of this example, suppose that 10 of the variables (the first 10, say) do not differ between the populations (but that at least one of the others does). Let us focus attention on the 10 that do not differ. Of course, we the experimenters do not know that the populations have identical means on these 10 variables. Thus, we might conduct 10 separate two-sample t-tests. To keep things simple, let us initially suppose that within each group the variables are uncorrelated. That means that the score on any one will not help us predict the score on any other. If we adopt the 5% significance level then we have a 5% chance of rejecting the null hypothesis on each of these variables (recall that, unknown to us, the null hypothesis is true for the 10 variables we are discussing). Conversely, on each of these variables we have a 95% chance of correctly concluding that there is no difference. But – and here comes the crunch – because the variables are independent, the probability of drawing this correct conclusion simultaneously on all 10 variables is only $(0.95)^{10} = 0.60$. This is a far cry from the 95% chance one might have imagined one had. It represents a 40% risk of incorrectly rejecting one or more of the null hypotheses.

In fact, of course, the 10 variables are unlikely to be independent, but are likely to be correlated in some way. This makes things worse because now we cannot calculate the probability of incorrectly rejecting one or more of the null hypotheses by the simple argument illustrated above, although it is true that 40% is an upper limit to the true probability.

Some readers might feel that there is no problem – tests on true sub-hypotheses will only be carried out when the true overall hypothesis is incorrectly rejected, and this will only have a 5% chance of occurring (if the 5% level is chosen). Thus true sub-hypotheses can only be incorrectly rejected at less than the 5% level overall. This is so – if the overall hypothesis is true. In the example above, however, the groups did differ on at least one of the 20 variables and the overall hypothesis was properly rejected; this simple argument thus breaks down. Further complications are introduced when we also start to examine possible *combinations* of variables and when more than two groups are involved.

Had the basic variables been the subject of prior hypotheses – had each been of particular interest in its own right – then, although the overall type I error probability would have been just as intractable, the problem may not have arisen. This is because the researcher would then wish to know about each variable separately and so, as we pointed out in Section 4.1, one might maintain that it would have been appropriate to test each variable at a significance level relating to it and it alone. In such a case, one would not be interested, necessarily, in making joint probability statements about the variables. In the present case, however, the individual variables are not of primary interest; they only become the subject of attention after the general hypothesis of identity of the two population means has been rejected. Thus it would not be appropriate to promote them each to the centre of the stage and test each of them at its own 5% level regardless of the overall level. We can describe the situation as being one of *exploratory* analysis – we are searching for variables or combinations of variables for which the null hypothesis will be rejected. Other terms to describe it are *post hoc* or *unplanned*.

Now that we have identified the problem and seen when it arises, we turn to ways to tackle it. A number of ways have been proposed, but the basic idea is to set a probability of incorrect rejection of the overall null hypothesis and to carry out the constituent tests at such levels that the overall level is not exceeded. We shall see examples of this below, but the reader should understand that fixing the overall level in this way means that the constituent tests might be carried out at a very extreme significance level (i.e. be very unlikely to reject a true null hypothesis). One cannot have something for nothing.

5.3 SOME BASIC IDEAS

When we explore a data set looking for models which fit, we might find more than one which seems to describe it adequately. That is, we might find alternative ways of describing the pattern of differences among the group means.

DEFINITION A comparison is *adequate* to describe the between-groups differences if there are no significant additional patterns of differences beyond those included in the comparison.

It will be clear from this that a complete comparison is always adequate. What we are after are adequate incomplete comparisons.

The reader may be familiar with rival adequate models from simple univariate regression analysis: if either of two predictor variables is by itself significant, yet in each case the other is not significant in addition to the first, then we have two adequate rival prediction models. In such cases both models can be reported on an equivalent footing, or the researcher can make a choice between them using external knowledge. Note that on the definition given above, the prediction model consisting of *both* variables is also adequate.

Any comparison other than a single contrast involves constituent comparisons. Put conversely, any incomplete comparison is a component of other comparisons. (It is at least a component of the complete comparison containing it, and perhaps other intermediate incomplete comparisons also contain it.) We can think of the null hypothesis that a particular comparison is zero as being implied by any wider null hypothesis relating to a comparison that contains it. This leads us to the following definition.

DEFINITION   A simultaneous test procedure is *coherent* if, when it rejects a null hypothesis that a comparison is zero, it also rejects all null hypotheses that comparisons containing it are zero.

Thus coherent test procedures maintain logical consistency: if the hypothesis that *A* equals *B* is rejected, so is the hypothesis that *A* and *B* and *C* are all equal. The first statement is contained in the second; if the first is rejected, so must be the second.

Another useful property is the reverse of coherence, concerning what an hypothesis implies about the component hypotheses it contains.

DEFINITION   A simultaneous test procedure is *consonant* if, when it rejects the null hypothesis relating to some comparison, then it rejects at least one of the null hypotheses relating to the component sub-comparisons.

This means that if any comparison is reported as being non-zero, then some contrast within that comparison will be identified as non-zero. It also means that if a multivariate test involving several variables has rejected the null hypothesis then a test on some particular weighted sum of those variables (i.e. on some single new derived variable) will also show significance.

Note that the fact that a test is consonant does not mean that it will be easy to find a significant sub-comparison or, indeed, that when found it will be of any interest. Because it involves a pattern of weights that can be anything at all, it may yield a result which cannot be usefully interpreted.

5.4 SEQUENTIAL TESTS

In Section 5.1 we referred to the nested sequence of tests arising from fitting polynomial components. Let us examine this example in more detail. It applies to both univariate and multivariate between-groups analysis.

We have a number of groups, defined by a single factor, such that the groups have some natural order (from low dose to high dose, or time 1 to time *t*, or low intensity of exposure to high intensity) and we wish to represent the pattern of group means in a convenient way. A common way is via polynomial trends across the groups: mean level, linear trend, quadratic trend, etc. For example, suppose we have three (equal size) groups of subjects, the first tested on day 1, the second on day 2 and the third on day 3. Then we might wish to know whether there is any change over time (is the mean level adequate to describe the overall pattern of means?) or if there is a uniform change over time (is the day 1 to day 2 change the same as the day 2 to day 3 change?). To answer these questions we conduct a sequence of (two) tests. First the difference of the day 1 to 2 change and the 2 to 3 change is examined (at level α_1, say). If the null hypothesis of no difference seems plausible then we assume the two changes are the same and test the common size of this change (at level α_2, say). The overall probability of a type I error in such a procedure is then

$$1 - (1 - \alpha_1)(1 - \alpha_2).$$

This generalizes in a straightforward way to more than two tests.

Using this formula we can work backwards, fixing a level for the overall probability and choosing the levels for each test so that the overall level is not exceeded.

The same principle can also be applied to examining sequences of variables. For example, if we had measures, some of which were known to be highly likely to distinguish between the groups and others of which we were less certain, then we might conduct a series of tests: for each variable, beginning with the least certain, we would use it to test the comparison, using as covariates the other variables which we were more confident distinguished between the groups. When such a procedure begins with all of the variables and sequentially deletes them in a pre-specified order (covarying out all those still remaining, except the one under test) it is called a *Roy–Bargmann step-down procedure*. Pre-specification of the order is a necessary requirement for the significance levels to be valid.

Display 18 Testing polynomial trends across groups

A researcher wished to explore in detail how length of practice of a certain kind affected the acquisition or problem-solving skills. Because it was felt that the testing procedure would interfere with the measure of length of time (the test involved further practice of a different nature) it was not adequate simply to test the same group at various times during practice. Instead four different groups were used,

Table X.18.1 Date for Display 18.

Group 1 (tested after 1 hour)
5, 6, 7, 6, 5, 7

Group 2 (tested after 2 hours)
6, 9, 12, 9, 10, 8

Group 3 (tested after 3 hours)
13, 13, 11, 7, 11

Group 4 (tested after 4 hours)
7, 15, 14, 14, 10

Table X.18.2 Anova table for polynomial trends across groups.

	Sum of squares	d.f.	Mean square	F	P
Constant	1 910.227	1	1 910.227	365.788	.000
Linear	111.282	1	111.282	21.309	.000
Quadratic	5.491	1	5.491	1.051	.319
Cubic	0.000	1	0.000	0.000	1.000
Within	94.000	18	5.222		

letting them practise for different lengths of time. The resulting data are given in Table X.18.1.

Table X.18.2 shows the analysis of variance results. In this table each effect is calculated assuming those physically below it are zero. The significance tests show that this is reasonable for the cubic and quadratic effects, and that we stop at linear. If each of these tests is conducted at the 5% level, the overall probability of a type I error is

$$1 - (1 - .05)^3 = 0.14$$

Conversely, if we wish an overall level of 5%, then if we conduct each test at the same level this common level must be 0.017.

5.5 GABRIEL'S METHOD

Gabriel's simultaneous test procedure is based on test statistic (4) of Section 4.3. It may be stated very succinctly: reject all those null hypotheses for which the relevant statistic exceeds a *common* critical value, this being the critical value for the overall hypothesis using all the variables simultaneously. Thus, to apply this method one first calculates the critical value for statistic (4) for the overall hypothesis and then compares the value of statistic (4) for each sub-test with this overall critical value.

Gabriel's method is coherent: if one hypothesis implies another, then rejection of the latter means the former must be rejected. It is also consonant. In fact this is only so because the set of hypotheses tested includes all possible between-groups contrasts and all possible linear combinations of variables. This is the point made at the end of Section 5.3. A consequence of the coherence is that the results of all the tests comprising the simultaneous test procedure can be easily summarized, either by listing the *minimal rejected hypotheses* (those which are rejected but do not imply any component rejected hypotheses) or by listing the *maximal accepted hypotheses* (those which are not rejected although any covering hypothesis implying them is rejected).

If the overall test is carried out at level α, giving a critical value $S(h_o, e_o; \alpha)$, where h_o and e_o are the hypothesis and error degrees of freedom respectively for the overall hypothesis, then the level γ for an individual test can be calculated using

$$S(h_o, e_o; \alpha) = S(h_i, e_i; \gamma)$$

where h_i and e_i are the hypothesis and error degrees of freedom for the individual test and γ is its individual significance level. Perhaps we should again add the remark that with many variables the individual sub-tests are very conservative. This is the price we pay for being able to control the overall error level.

Display 19 An example of Gabriel's test

A researcher, interested in the three creativity variables described in Display 1, is unwilling to state, a priori, which particular patterns of differences between group means are of concern. Instead it is required to be able to explore all possible patterns of differences. Moreover, should a test on all three variables simultaneously prove significant, then the researcher will wish to follow it up to see if some one or more individual variables account for the overall difference. It is therefore decided that a simultaneous test procedure is needed to provide overall protection at a specified significance level, and that Gabriel's test is appropriate. The data are given in Table X.19.1.

The 5% significance level for the overall test, involving three groups and three variables, requires a Roy's maximum root statistic; let us call it R (i.e. $\lambda_1/(1 + \lambda_1)$, in the terminology of Section 4.3) equal to 0.535. Thus overall protection at the 5% level is given if we compare any subtest with this value.

Table X.19.2 shows the results, these being Roy's maximum root statistic (the values of R) for all subsets of two or more groups and all subsets of variables. We see immediately that none achieves the required level. Of course, protecting ourselves at the 5% level means that each subset is made at a stricter level. This level can be calculated as follows:

1. For univariate one-degree-of-freedom tests the value of the F-statistic is calculated as

$$F(1, n_e) = n_e \lambda_1 \quad \text{or} \quad F(1, n_e) = n_e R/(1 - R)$$

   where λ_1 is the largest eigenvalue or R is Roy's maximum root statistic, and n_e is the error degrees of freedom.
2. Using this, given a value for R, we can calculate the F-statistic $F(1, n_e)$ and hence, via F-tables, the associated significance level. In our case the required value for R of .535 gives

$$F(1,15) = .535 \times 15/(1 - .535) = 17.26$$

3. F-tables show this corresponds to a significance level $P < .005$. That is, if we insist on overall protection at the 5% level then each univariate single-degree-of-freedom test (e.g. between just two groups) is being conducted at the 0.5% level.

   We might feel that this level of significance is too extreme and wish to relax it. We can do this by beginning at the other end of the problem and specifying at what level

Sequential and simultaneous test procedures

Table X.19.1 Data for Gabriel's test.

Group 1			Group 2			Group 3		
C1	C2	C3	C1	C2	C3	C1	C2	C3
5	11	14	1	20	16	3	9	21
10	9	15	1	15	16	11	9	21
16	8	14	3	13	13	16	9	18
17	7	15	7	12	20			
2	15	17	9	10	21			
			15	9	15			
			26	8	20			
			11	8	11			
			14	8	10			
			23	7	16			

we want each univariate single-degree-of-freedom test to be conducted. In principle one can then work the above calculation backwards to find the overall protection level. In practice this is seldom feasible because of the limitations of the available tables and charts, but we shall illustrate how it is done.

1. Suppose, for illustration, that we require each univariate single-degree-of-freedom test to be conducted at the 5% level. From F-tables we see that this corresponds to $F(1,15) = 4.54$.
2. The above relationships between $F(1,n_e)$ and R then give

   $$4.54 = 15 \times R/(1 - R)$$

   from which $R = 0.232$.
3. It is at this point that the phrase 'in principle' above comes into effect. If one had sufficiently detailed charts of R one could then relate this value of R, with the number of groups and variables of the overall test, to an overall significance level.

The bold figures in Table X.19.2 are those showing significance when each univariate single degree of freedom test has the 5% level (i.e. those larger than .232). We can see the coherence principle in action here: (a) for any particular set of variables, if the test on the difference between a pair of groups is significant then so is the test on all three groups; and (b) if any set of groups differs significantly on some set of variables then they differ significantly on any set which includes that set.

The tests shown in Table X.19.2 are not all the possible tests. They do not show tests on all possible weighted sums of variables (of which there are an infinite number). This explains the lack of consonance manifest in the table: groups G2 and G3 differ significantly on the pair of variables C2 and C3, but not on either separately. Had all possible tests been conducted then some particular weighted sum of the variables C2 and C3 would show significance.

The results of this simultaneous test procedure can be summarized in the list of

Table X.19.2 Roy's maximum root statistic for all subsets of two or more groups and all subsets of variables in the data of Table X.19.1.

Variables	Groups			
	(G1, G2, G3)	(G1, G2)	(G1, G3)	(G2, G3)
C1, C2, C3	**.401**	.086	**.310**	**.397**
C1, C2	.182	.084	.026	.157
C1, C3	**.272**	.018	**.251**	.227
C2, C3	**.309**	.027	**.273**	**.281**
C1	.005	.004	.000	.002
C2	.051	.017	.010	.046
C3	**.271**	.015	**.251**	.226

Table X.19.3 Summary of the significant differences between groups given in Table X.19.2.

Minimal rejected hypotheses	Maximal accepted hypotheses		
(G1, G3) on C3	(G1, G2, G3)	on	C1, C2
(G2, G3) on C2, C3	(G1, G2)	on	C1, C2, C3
	(G2, G3)	on	C1, C3

minimal rejected hypotheses shown in Table X.19.3. Using the coherence property, these significant differences imply the existence of all the other significant differences found by the procedure (those in bold in Table X.19.2). The hypotheses in the list of maximal accepted hypotheses, in a similar fashion, contain only the accepted hypotheses in Table X.19.2.

One final point of detail: in deriving the test statistics in Table X.19.2 it is important to note that those for subsets of groups (incomplete comparisons) are derived by testing those incomplete comparisons as part of complete comparisons – that is, without restricting any other comparisons to zero. They are not obtained simply by selecting only that subset of groups for the analysis, which would, among other things, lead to different error sums of products matrices for the analysis of each subset.

5.6 McKAY'S METHOD

This method was introduced for multiple group discriminant analysis, but the principle is general. The question addressed is whether a subset of the variables provides essentially the same information on a between-groups comparison as does the entire set – that is, is the subset adequate to account for the comparison? The entire set is divided into two subsets: a set of covariates which will be assessed for adequacy; and the remainder, to be used as response ('dependent') variables. The first step is to examine the comparison on all variables simultaneously, rejecting the null hypothesis if the test statistic (McKay prefers criterion (3) of Section 4.3) is greater than the critical value for the chosen level. If the hypothesis of a zero comparison is not rejected then the process is concluded: no differences have been found. If this overall hypothesis *is* rejected then the search for adequate subsets proceeds by testing the hypothesis that the subset of covariates is adequate and that the response variables add nothing to the comparison. The corresponding statistic for the response variables adjusted for the chosen covariates is tested against the *same* critical value. If the set of covariates is inadequate the null hypothesis on the response variables will be rejected.

The level at which an individual test is being made under this procedure can be deduced in a way analogous to that described above for Gabriel's method. If a subset of size p from the d original variables is being examined to see if it provides essentially the same separation on the comparison, the significance level γ at which the test is actually carried out is computed from

$$S(h_o, e_o; \alpha) = S(h_o, e_o - p; \gamma)$$

As the size of the subset of covariates being examined increases, so the tests become more conservative.

5.7 MORE ON CHOICE OF TEST STATISTIC

In Section 4.4 we discussed choosing which test criterion to use in multivariate analysis. In that discussion we noted how overall power and robustness influence the relative performance of various criteria. However, the overall test is not the end of the matter. We must also consider the performance and properties of the tests when used in simultaneous test procedures as sequels to the intrinsically multivariate test. Gabriel notes that statistic (4) of Section 4.3 rejects at least every single contrast hypothesis rejected by any other simultaneous test procedure at the same level and so is the most *resolvent* simultaneous test procedure. Moreover, for (4), if the overall test is significant then it is always possible to find at least one contrast within the simultaneous test procedure which is significant. This is not the case for (1). Gabriel thus prefers (4) to the other statistics.

The choice between criteria (1) and (4) has been further explored by Kevin Bird and Dusan Hadzi-Pavlovic, who present a more extensive discussion. Our recommendations are as follows: if the researcher has confidence that the multivariate analysis of variance assumptions are true then (4) should be used. Otherwise, and this is perhaps the more common situation, (1) should be used except in the special but common case of all follow-up tests referring only to individual variables (and not to combinations of variables). In this case a more powerful set of tests results if the multivariate simultaneous test procedures above are replaced by univariate simultaneous test procedures (e.g. Scheffé's method), each with nominal error rate set at B/d, where B is the overall rate and d is the number of variables.

The reader should note, however, that the above are only recommendations, and until further research is done and more comparative experience accumulated one is unlikely to be criticized for poor choice of multivariate test statistic.

5.8 VARIABLE SELECTION

In the above we have examined ways in which subsets of variables, as well as individual variables, could be tested for their contribution or adequacy in explaining effects, without the large number of tests compromising the overall significance level. It remains to consider how to choose the subsets to be examined.

Often the question can be answered in a straightforward way: we simply examine all possible subsets of variables. However, with d variables there are $(2^d - 1)$ subsets of variables. So, with 10 variables, there are over 1000 subsets and with 20 variables there are over a million subsets. If each of these is to be examined then we will be in for a long wait. The property of coherence possessed by the methods of Gabriel and McKay aids the search (e.g. if a particular set of variables does not attain significance, then no subset of this set will) and other methods to accelerate the search have been developed, many in the context of discriminant analysis.

One common approach is to compare the standardized coefficients of the variables on the canonical variates. A variable with a large (positive or negative) coefficient is considered important. The standardization is, of course, essential, so that the comparison is not affected by an arbitrary choice of units. As we have already noted, unless the hypothesis involves only a single contrast there will be more than one canonical variate, so that in examining each variable more than one coefficient must be taken into account. A reasonable approach seems to be a sequential method, eliminating one at a time those variables which appear unimportant. After each elimination recalculate the canonical variates and, for each variable remaining, examine the sizes of the coefficient on those canonical variables which have canonical values sufficiently large that these variates are considered important. The stepwise nature of this approach is important because the values of the coefficients for each variable can depend on which other variables are included. (This is also true in regression analysis. If the variables were uncorrelated then the problem would not arise. The effect becomes greater the larger the extent of correlation.)

Another very important class of stepwise methods is based on testing to see if a variable adds anything to the hypothesis sum of squares beyond that due to variables already chosen. This can be done either *forward*, in which variables are added at each step, or *backward*, beginning with all the variables and sequentially eliminating them. If the order of examination of the variables is specified beforehand then Roy–Bargmann tests, as described in Section 5.4, can be applied. However, a common practice (in discriminant analysis programs, at least) is to let the data determine the order. For example, in a forward process one might choose to add that variable for which the F-statistic (of the test of additional information beyond that due to already chosen variables) was greatest. If the F-statistic is simply regarded as a measure of additional information, then this is perfectly legitimate. However, the nominal significance levels associated with such an F value are invalid – because at each step the chosen variable is the one with the *greatest F* value. We remarked on this point at the end of Section 5.4. Some programs print out these invalid significance levels.

Forward and backward stepwise methods as described above make feasible the search through subsets of a large set of variables by restricting the search. There is no guarantee that they will find the 'best' set (some authors dispute that such a term has general meaning). One of their weaknesses is that by examining variables one at a time, they might miss variables which are very important when taken together. This has led to the development of methods which examine small sets of variables (e.g. pairs) at each step rather than single variables. Of course, there is a cost in terms of computational effort.

Taken by itself, a disadvantage of the forward method is that a variable, once included, cannot be eliminated even though it might subsequently become superfluous due to later inclusion of other variables. This point has led to the development of methods combining forward and backward ideas in which attempts are alternately made to add useful new variables and delete superfluous ones. Of course, the comments made above about the lack of an adequate theory of significance testing apply for all such methods.

Because of the property, noted above, that stepwise methods render the search for subsets feasible by restricting it in certain ways, attention has also turned to methods for generating all possible subsets. For example, C. P McCabe has developed an efficient computer algorithm for computing statistic (2) for all possible subsets of a given set of variables. Other methods are based on the fact that certain of the test statistics are monotonically related to the number of variables in the subset. In such a situation *branch and bound* procedures can be applied. In all such cases, of course, simultaneous test procedures should b used to avoid gross inflation of the overall type I error rate.

5.9 CONFIDENCE INTERVALS

The reader will be aware of the duality between significance testing and confidence intervals from univariate studies. Thus, it will come as no surpris to learn that as an alternative (or supplement) to simultaneous test procedure one can set up simultaneous confidence limits. That is, for any given hypothes one can calculate confidence bounds such that, for any and all such hypotheses one cares to examine in an experiment, there is at least a 95% chance (say) that the interval covers the true value. However, since calculating these intervals requires a little matrix algebra, it would be beyond the spirit of this book to include the details. References are given below.

5.10 FURTHER READING

An overall introduction to simultaneous test procedures, both univariate and multivariate, is given by Miller (1981). Gabriel (1968) is an important paper describing simultaneous test procedures using the largest root statistic (4), and is to be recommended. Roy (1958) and Roy and Bargmann (1958) describe 'step-down tests', and McKay (1977) introduces the 'additional information test' outlined in Section 4.3.

Turning to variable selection methods, McKay and Campbell (1982) provide an excellent tutorial overview; McCabe (1975) presents an algorithm for exploring all possible subsets using the likelihood ratio statistic (2); and Hand (1981) describes the branch and bound method.

Descriptions of how to derive simultaneous confidence intervals may be found in Morrison (1967, p.182); Miller (1981, p.202); Gabriel (1968, p.498); and Srivastava and Carter (1983, p.98), among other publications. Aitkin (1978) describes simultaneous test procedures for sets of contrasts in multi-factor univariate analyses.

PART 2————The practice

Alcohol relapse ——————————— **Study A**

The first study in this series of examples is an investigation into the causes of relapse among alcoholics, conducted at the Addiction Research Unit, London, by Dr G. K. Litman and Dr A. N. Oppenheim, and our thanks are extended to them and their co-workers for allowing us to use their data. We begin the analysis with a simple univariate analysis of variance, in order to illustrate and help fix the ideas introduced in Chapter 2. Having introduced the principal questions under investigation in this more familiar setting of simple univariate analysis, in Section A.2 we proceed to a multivariate analysis of variance, using a parallel set of data taken from the same study.

The study consists of a sample survey of people presenting themselves for treatment of alcoholism and heavy drinking problems at several hospitals and other related agencies. The data were collected over a one-year period, with a view to a long-term follow-up of all those participating. The data in the present analysis consist of information taken at the point of the patient's admission into treatment. A central interest in the study was to inquire how vulnerable people with drinking problems are to relapse, and in what way this vulnerability is perceived. The measures with which we are here concerned were made with a 'Relapse Precipitants Inventory' (RPI) which all of the participants in the study were asked to fill out. This questionnaire was developed in a previous survey and listed a variety of situations which were likely cues for drinking. Respondents scored these items according to how dangerous they found them in precipitating a relapse into heavy drinking. The sum of these item scores provided a total score for the entire inventory and it is this total which we now analyse.

The sample was drawn from a year's total consecutive admissions to treatment for alcoholism in the hospitals and agencies referred to above. These consecutive admissions were subjected to various exclusion criteria for a variety of administrative and ethical reasons. This left over the one-year period 256 cases, of which 5 failed to give full information on the Relapse Precipitants Inventory. The study data thus comprise the remaining 251 cases, of which roughly 25% are female and 75% are male. The main interest of the study concentrated on future relapse or non-relapse, but it was also important to assess the present position of the subjects in the study. In particular, information was collected on their personal histories and previous troubles with heavy drinking. This enabled the

Alcohol relapse

researchers to characterize the subjects by the number of previous relapses they had experienced, prior to their current heavy drinking period and consequent intake into the study. It is therefore possible to inspect the relationship between previous experience of relapse and present perceived vulnerability to relapse.

Although it would be possible to look for a linear relationship between these two measures by means of a regression or correlation analysis, it was thought that the spread of experience of previous relapses would not make for a simple linear relationship throughout the range. Furthermore, this particular item of information is not easy to define accurately and consequently there are fairly arbitrary elements introduced into its measurement. It was therefore decided to categorize the people in the sample into three crude groups according to the rough count of previous relapses. These groups were defined as

- those never having previously experienced relapse after trying to give up heavy drinking;
- those who claimed to have relapsed, but no more than two or three times before;
- those who had a longer history of relapse of four or more times.

These three groups are referred to as 'new', 'recent' and 'long-standing'. The numbers in each group are shown in Table A.1, along with the mean score for the group on the RPI. The score is a summation of the scores on 25 individual items, with a maximum score of 75.

Our first analysis is therefore a univariate analysis of variance with one factor at three levels representing the experience of the subject. The intention is to inspect the influence of the 'experience' factor on the response measure, vulnerability. As described in Section 2.1, in a simple analysis of variance we usually define group effects, that is, the effects of the factor at its three levels, as the difference between the individual group mean and the overall mean of the three groups. Such effects are represented in terms of a set of contrasts by the following:

$$
\begin{array}{ccc}
1 - 1/3 & -1/3 & -1/3 \\
-1/3 & 1 - 1/3 & -1/3 \\
-1/3 & -1/3 & 1 - 1/3
\end{array}
$$

Table A.1 Numbers in cells and means on the RPI score for the three levels of the factor 'experience'.

Factor level	No. in cell	Observed mean
Group 1 (long-standing)	125	40.00
Group 2 (recent)	90	34.31
Group 3 (new)	36	31.78
Overall	251	36.78

Table A.2 Possible alternative sets of contrasts constituting the complete comparison of the three groups shown in Table A.1 and the estimates of the effects they define.

Group 1	Group 2	Group 3	
Obs. mean scores:			
40.00	34.31	31.78	
Grand mean:			Corresponding est'd effect:
1/3	1/3	1/3	35.363
Standard contrasts:			Corresponding est'd effects:
1 − 1/3	− 1/3	− 1/3	4.637
− 1/3	1 − 1/3	− 1/3	−1.052
− 1/3	− 1/3	1 − 1/3	−3.585
Pairwise differences contrasts:			Corresponding est'd effects:
1	−1	0	5.687
0	1	−1	2.533
1	0	−1	8.222
Polynomial contrasts:			Corresponding est'd effects:
−1	0	1	−8.222
−1	2	−1	−3.156
Helmert contrasts:			Corresponding est'd effects:
1	−1/2	−1/2	6.956
0	1	−1	2.533

These three contrasts (or indeed, any two of them), it can easily be shown, constitute a complete comparison for the factor, as described in Section 2.2. The estimated values of the effects are given by the values of these three contrasts when applied to the sample means and these are shown in Table A.2. Note that these effects, which compare the group mean with the overall mean, are simply multiples of the effects which compare each mean with the average of the other two groups. This is easy to see from the following three contrasts, which are simply 1½ times the previous set:

1	− 1/2	− 1/2
− 1/2	1	− 1/2
− 1/2	− 1/2	1

Again as was indicated in Section 2.2, this implies that the effects are equivalent, in that each measures the same set of relationships.

Alternatively, we can define the effects as they were first described in Section 2.2, as all possible pairwise differences between the group means. In this particular case, of three groups only, we have three pairwise comparisons and these are also given Table A.2, along with the estimated values of the effects defined in this manner. These estimated values in the complete comparison are again, of course, just the value of the contrasts applied to the observed group means. The two sets are alternative representations for the complete comparison of the levels of the factor 'experience'.

In this example the levels of the factor have an order to them, in that they progress from many relapses in the 'long-standing' group to none in the 'new' group. Under such circumstances it is often desirable to look for trends across the three groups and in particular the polynomial trends, linear and quadratic. Effects defined this way, as trends, are also shown in Table A.2. It can be readily verified that these two contrasts also provide a complete comparison for the factor. They constitute one particular summary of all the pairwise differences described previously (see Section 2.2).

Although a case can be made for using the two polynomial contrasts to represent the ordered effects in this study, it was felt on balance that they were not an interesting way of describing the differences between the levels of the factor. Prior considerations led the researchers to the view that the first group, that of long-standing relapsing heavy drinkers, would be considerably different from the two other groups which had only new or recent histories of problematic drinking. These two groups, it was felt, would show no great differences from each other, whereas their joint differences from the group which contained a large number of very long-standing problem drinkers would be quite marked. It was therefore decided that the effects of primary interest for this factor are the two defined by the two contrasts shown in the last part of Table A.2. Contrast one represents the marked difference between the first group and the other two, and contrast two represents the difference between these last two groups, 'new' and 'recent'. The table also gives the value of these contrasts calculated on the observed group means. Note that these two contrasts (which, incidentally, are the Helmert contrasts referred to in Section 2.11) provide yet another alternative representation of the complete comparison for the three groups, a fact which is again easily verified. It is with this last set of contrasts that the analysis shall concern itself. It is important to appreciate that these contrasts were decided upon simply from a priori considerations and interests, without reference to the observed data. As described in Section 2.2, they are conceived as being relationships between the population means of the three groups; when applied to the sample means they produce the estimates of these population effects which are given alongside them in Table A.2.

The estimates for these two contrasts were produced with a standard computer program, although a program is scarcely needed for such simple computations. However, computing the sums of squares for the related tests of significance is a different matter. Table A.3 is the standard analysis of variance table, although some readers may be used to one which omits the sum of squares due to the grand mean. It shows the familiar sums of squares attributable to the factor calculated with such a program. Note that this overall sum of squares for

Table A.3 Univariate analysis of variance of the RPI score for the data shown in Table A.1.

(i) Analysis of variance for the complete comparison.

Source	d.f.	SS	F-statistic	(sig-level)
Within groups	248	75 345.5		
Grand mean	1	339 561.0		
Between groups	2	2 745.4	4.518	(.012)

(ii) Analysis of variance for the partitioned between-groups degrees of freedom, showing the 'unique' contribution of each to the complete comparison.

Source	d.f.	SS	F-statistic	(sig-level)
Within groups	248	75 345.5		
Grand mean	1	339 561.0		
1st Helmert (1 vs. 2, 3)	1	2 729.9	8.985	(.003)
2nd Helmert (2 vs. 3)	1	165.0	0.543	(.462)

the factor – that is, for the complete comparison of all levels – would be the same value no matter which of the sets of contrasts were used to represent the complete comparison. The within-group sums of squares are also shown in the table. The usual F-test would compare the mean square associated with the sums of squares due to the comparison with the mean square for error in the familiar way. This F-statistic would be used to test the overall null hypothesis that there are no differences between any groups (see Section 4.2). Such a test of hypothesis would be the appropriate one when searching for any, or general, differences between the three groups. In this particular experiment, however, the overall null hypothesis of no differences is not the one to test, for the researchers had in mind two particular hypotheses which they desired to test separately. These two hypotheses are the ones represented by the contrasts themselves, namely that group 3 is equal to group 2 and that group 1 is different from this average. There is no particular reason to construct a simultaneous test of these two distinct hypotheses.

Thus the second half of Table A.3 shows the tests of the hypotheses in which we are actually interested. The two separate contrasts are shown as partitioned sources of variation, each with one degree of freedom. The sum of squares opposite each entry is the sum of squares attributable to the contrast when allowing for everything else in the complete comparison – in this case, that means just the other contrast. These are referred to as the uniquely attributable sums of squares for the contrasts, or more simply as the 'unique' sums of squares. They measure the need to include the particular contrast in the comparison, that is, they are each used to test the null hypothesis that the corresponding population contrast value is zero. We note in passing that the two sums of squares do not add up to the overall sum of squares for the complete comparison (in spite of the two contrasts 'appearing' orthogonal) because of the imbalance in the design (see Section 2.6).

The two relevant F-statistics are quoted alongside the corresponding unique sums of squares in Table A.3 and are, as usual, the ratio of the mean square for the hypothesis to the mean square for error. Each hypothesis is tested separately at the 1% level, this high level being chosen simply because a lot of data are available (251 observations in all), allowing stronger conclusions to be drawn. The first test of hypothesis – that the first Helmert contrast (comparing the 'long-standing' group with other two) is zero – is rejected, showing that the 'recent' and 'new' group average is significantly different from the 'long-standing' group. We can see from the bottom line of the table that the null hypothesis that groups 2 ('recent') and 3 ('new') are the same is not rejected. Thus there is no evidence for this effect being anything other than zero. The contrast can be omitted from the comparison and the remaining partial comparison – that is, the contrast representing the two groups' joint difference from the 'long-standing' group – is adequate to explain the observed data.

Under these circumstances of groups 2 and 3 being equal, we might wish to proceed to estimate the incomplete comparison which comprises the adequate model, that is, the single contrast representing this joint difference between the 'long-standing' group and the other two. Recall from Section 2.3 that this is equivalent to pooling groups 2 and 3 for the purposes of estimating their average difference from the 'long-standing' group. The value for the contrast estimated as an incomplete comparison, with group 2 restricted equal to group 3 by the omission of the corresponding contrast from the comparison, is shown in Table A.4(i) as 6.413. Note that the contrast coefficients which are shown as

$$1 \qquad -1/2 \qquad -1/2$$

are the 'conceptual' weights on the population group means referred to in Section 2.3; they are not the efficiently reweighted coefficients actually applied to the observed group means to produce the estimated value of the effect.

Most computer programs provide the facility for reconstructing, or 'predicting' as it is sometimes termed, the observed group means from the effects in a diminished model of this sort. Thus the reconstructed means shown in Table A.4(i) are formed from the single significant group effect and the overall mean. These are the estimates of the group means made under the assumption that the 'recent' and 'new' groups have equal values.

We can also test in this diminished model whether the size of the effect is non-zero. The null hypothesis is that the population value of the first Helmert contrast is zero, and the effect therefore superfluous, even when it is the only one in the design. Thus we need to use the sum of squares for the partial comparison

Table A.4 (i) Estimation of effects under the assumption that 2nd Helmert (1 vs. 3) has population value of zero, along with estimated group means under this assumption.

	Conceptual coefficients			Est'd effect
Grand mean:				
	1/3	1/3	1/3	35.725
1st Helmert:				
	1	−1/2	−1/2	6.413
Est'd group means:				
	Grp 1	Grp 2	Grp 3	
	40.00	33.587	33.587	

Table A.4 (ii) Analysis of variance for the minimal adequate model (i.e. 2nd Helmert contrast assumed zero).

Source	d.f.	SS	F-statistic	(sig-level)
Within groups	248	75 345.5		
Grand mean	1	339 561.0		
1st Helmert (1 vs. 2, 3)	1	2 580.4	8.493	(.004)

(i.e. for just the remaining contrast) in an *F*-test. The result, shown in Table A.4(ii), is a highly significant one and we reject the null hypothesis: there is such an effect, and so it constitutes a minimal adequate explanation of the data.

Before turning to the multivariate analysis, we shall briefly recapitulate on the procedures used here. It was decided *a priori* to represent the complete comparison between groups by the two Helmert contrasts. Each of these was to be the subject of a separate test of hypothesis that the associated relationship between group means was zero. Only one of these null hypotheses was rejected, that which declared a significant difference between the first group and the other two. This contrast, or incomplete comparison, thus constituted an adequate explanation of the data; and the estimated value of this incomplete comparison was recomputed, by using the information that the mean of group 2 is equal to the mean of group 3.

A.2

We turn now to a multivariate analysis of variance on a set of variables from this same study. It will be recalled that the RPI was developed from work on a previous survey. This work had identified three areas of vulnerability to relapse,

Alcohol relapse

which were incorporated in the 'Relapse Precipitants' questionnaire. These three areas gave rise to three separate scores, one for each, as follows:

(UM) unpleasant mood states; for example, depression;
(ES) euphoric states and related situations; for example, celebrations and parties; and
(LV) an area designated as lessened vigilance; for instance, a temptation to believe that one or two drinks would cause no problem.

We shall now parallel the analysis of the preceding section, but using these three separate scores to characterize the individuals in the three groups, 'long-standing', 'recent' and 'new'. In fact, this analysis was the more important analysis in the research, the univariate analysis of the whole RPI being presented here rather for the purpose of giving an initial description of the questions under study. Accordingly we shall, in analysing these data, ignore the fact that we have already univariate results for a related problem. (The raw data for this study are given in Table A.11.)

We thus have for each subject three response variables representing three vulnerability scores; between subjects the design consists, as before, of one factor ('experience') at three levels. The questions in this analysis concerning these three levels of experience are the same as those in the preceding analysis; that is, the two separate questions of firstly, whether 'recent' (group 2) and 'new' (group 3) are different; and secondly, whether 'long-standing' (group 1) is different from their average. In this analysis, however, we are concerned with the groups being equal or different across the profile of scores representing the three areas of vulnerability. We are therefore interested in whether the group mean vectors (not just univariate group means) differ. There are no particular relationships postulated between these three variables so the questions are general ones, asking simply whether the variables as a set can be used to identify the between-groups relationships described above (see Section 3.2). Table A.5 shows the mean vectors for the three groups under study.

As in the univariate example above, we begin the analysis by estimating the effects in the complete comparison, represented by the two Helmert contrasts

Table A.5 Numbers in cells and mean vectors on the three RPI scales for the three levels of the factor 'experience'.

Factor level	No. in cell	Observed mean vector		
		(UM)	*(ES)*	*(LV)*
Group 1 (long-standing)	125	21.60	11.98	6.42
Group 2 (recent)	90	18.24	10.68	5.39
Group 3 (new)	36	17.33	9.94	4.50
Overall	251	19.78	11.22	5.77

previously described. These are shown in Table A.6 and we note that for each effect we now produce a vector of values, one for each variable. Along with the effects relating to the Helmert contrasts is shown the grand mean vector (the overall level for each variable). This differs from the simple sample mean shown at the foot of Table A.5, because it is fitted in conjunction with the contrasts and therefore uses the conceptual coefficients shown directly. We now proceed to test the hypotheses that the Helmert effects are zero in the population. As in our univariate case, the research arguments lead us to test the two hypotheses separately.

Table A.6 The Helmert contrasts for the complete comparison of the three groups shown in Table A.5 and the estimates of the effects they define.

Grp 1	Grp 2	Grp 3	(UM)	(ES)	(LV)
Grand mean:			Corresponding effect vector:		
1/3	1/3	1/3	[19.06	10.87	5.43]
Helmert contrasts:			Corresponding effect vector:		
1	−1/2	−1/2	[3.81	1.47	1.47]
0	1	−1	[0.91	0.73	0.89]

Table A.7 is a parallel to Table A.3, but shows matrices of sums of products whereas Table A.3 shows only sums of squares. For each response variable in the multivariate analysis there is a column and a row in the sums of products matrix, and the diagonal entry in the matrix corresponds to the sums of squares that would be used for that variable in a straightforward univariate analysis, as was described in Section 3.1. Here, however, we are interested in all three variables simultaneously and in the relationships between them. In other words, we compare the whole matrix of sums of products for the hypothesis with the whole matrix of sums of products for the error term.

The topmost entry is the within-groups sums of products matrix **E** that we shall use for the error term in the significance tests. The second entry is the hypothesis matrix for the grand mean, which we shall ignore since we do not wish to test whether this is zero in the present example. The next entry, with two degrees of freedom, is the hypothesis matrix relevant to the complete comparison between groups. It is included here only to make the parallel with Table A.3 exact and we shall ignore it since, as we argued in the previous analysis, we do not wish to test the compound hypothesis that there are no group differences. We wish only to perform tests on the two individual hypothesis matrices shown for the two separate hypotheses relating to the first Helmert contrast and the second Helmert contrast. The sums of products matrices for these partitioned sources are given in the lower part of the table. To test the null hypothesis that group 1 equals the average of groups 2 and 3 in their mean vectors, we compute the

Alcohol relapse

Table A.7 Multivariate analysis of variance of the three RPI scales and the factor 'experience' shown in Table A.5.

Source	d.f.	SP matrix			Pillai statistic
		(UM)	(ES)	(LV)	
Within groups	248	30 637	—	—	
(**E** matrix)		10 152	10 440	—	—
		3 878	2 251	1 707	
Grand mean	1	98 251	—	—	
(**H** matrix)		55 733	31 615	—	—
		28 688	12 262	8 364	
Between groups	2	841.76	—	—	.071
(**H** matrix)		361.09	342.07	—	(sig .006)
		311.49	138.61	123.30	
1st Helmert	1	819.57	—	—	.068
(**H** matrix)		359.75	157.91	—	(sig .001)
		316.45	138.91	122.19	
2nd Helmert	1	21.35	—	—	.013
(**H** matrix)		17.18	13.83	—	(sig .347)
		20.83	16.76	20.32	

matrix ratio $\mathbf{H \cdot E^{-1}}$, using the hypothesis matrix $\mathbf{H}$ relating to the first Helmert contrast. The three eigenvalues of this matrix ratio (see Section 4.3) are

0.0734   0.0   0.0

The test statistic we shall use is the Pillai–Bartlett statistic (see Section 4.3) and this is shown in Table A.7 opposite the appropriate entry. This statistic is significant and we reject the null hypothesis, concluding that there is a difference between the mean vector of group 1 and the mean vector of groups 2 and 3. The second hypothesis is tested in an identical way, using the hypothesis sums of products matrix relevant to the second contrast. The eigenvalues of this matrix ratio are

0.0133   0.0   0.0

and the derived Pillai–Bartlett criterion is shown in the table. This does not yield a significant value. As in the univariate analysis above, we are testing to see whether the contrast representing the specified relationship needs to be included in the overall comparison in order to give an adequate description of the data. Our conclusion is that the first contrast alone is adequate.

Having decided that the first Helmert contrast is an adequate model for the data by itself, we can re-estimate its value on the three response scales using the

Table A.8 (i) Estimation of effects under the assumption that the 2nd Helmert contrast (2 vs. 3) has population value of zero, along with estimated group means under this assumption.

Conceptual coefficients			Estimated effects		
Grp 1	Grp 2	Grp 3	(UM)	(ES)	(LV)
Grand mean:			Corresponding effect vector:		
1/3	1/3	1/3	[19.19	10.97	5.56]
1st Helmert:			Corresponding effect vector:		
1	−1/2	−1/2	[3.62	1.52	1.28]

Estimated group mean vectors	(UM)	(ES)	(LV)
Grp 1 'long-standing'	[21.60	11.98	6.42]
Grp 2 'recent'	[17.98	10.47	5.14]
Grp 3 'new'	[17.98	10.47	5.14]

Table A.8 (ii) Multivariate analysis of variance for the minimal adequate model (i.e. 2nd Helmert contrast assumed zero).

Source	d.f.	SP matrix			Pillai statistic
		(UM)	(ES)	(LV)	
Within groups	248	30 637	—	—	
(**E** matrix)		10 152	10 440	—	—
		3 878	2 251	1 707	
Grand mean	1	98 251	—	—	
(**H** matrix)		55 733	31 615	—	—
		28 688	12 262	8 364	
1st Helmert	1	820.41	—	—	.059
(**H** matrix)		343.91	144.16	—	(sig .002)
		290.67	121.85	102.98	

information that groups 2 and 3 have equal mean vectors. The results are shown in Table A.8(i), with the reconstructed values of the group means estimated under this restriction. We can see that in this particular example the estimate of the effect is only slightly changed. However, in general such estimates are more efficient, given that the restriction is appropriate.

This procedure is sometimes referred to as 'trimming the analysis' of its superfluous effects and we can proceed to look at the relationships of the variables under these restrictions. Having shown that the three variables UM, ES and LV as a set distinguish the 'long-standing' group from the others, we may use the canonical representation of the set to highlight the structure of this

107

discriminant ability. There is only one canonical variable for this comparison, since it comprises only one degree of freedom (see Section 3.10). Calculations using a computer program give the optimum combination of the three responses as being obtained by using the respective weights

0.247   – 0.172   0.939

on the standardized variables – that is, on each response after dividing it by its (within-group) standard deviation. This standardization allows us to compare the relative importance of the responses in contributing to the canonical variable, even though they may be measured on very different scales. It can be seen that the primary weighting is on the third response, LV, and we conclude that this measure alone will give almost as good a discrimination.

A.3

We turn now to a second way of analysing these multivariate data. We seek to answer a different question in this analysis, one which was not part of the initial research aims but which is included here to demonstrate the relationship between the two multivariate analyses. The question we will ask is 'has anything been gained in analysing the differences between the groups by using three separate scales for vulnerability, instead of just a simple overall level?'

The analysis of Section A.2 looked at the profiles of the groups on the three variables without making any specific hypotheses about the sort of relationships that might occur between these measures. It is an example of what in Sections 3.2 and 3.8 is referred to as an unstructured question about the variables. We have now effectively structured the question by asking about specific, predetermined combinations of the variables, namely their overall average (the general level of response) and the difference of each of them from this overall average. We therefore consider them explicitly as three separate levels of a single between-variables factor which we shall call 'type'. Thus we are in a position to use the same structures and transformations described for the between-subjects design in Section 2.15, separating an overall mean level of response from a complete comparison between the variables.

Section 3.4 explained the need to have some sort of commensuration across the variables if we are going to use specified linear combinations of them and be able to interpret the results. In the preceding profile analysis it was not necessary, since no such transformations were specified. In this present context we note that each variable is a sum of several items in the RPI (each item being scored in an identical way), so an overall total of the variables is interpreted as the extent to which a subject is measured as vulnerable to relapse by the inventory. Differences between the scales are similarly differences in these extents. We will expect to find large differences between the scales, since the number of items in each is not the same, and such a finding will not contribute to our understanding of the data. None the less, whether these differences are the same for each group or larger in some than others is a useful question to ask.

We shall define our between-variable factor effects as the following set of combinations:

V0:	1	1	1
V1:	1	-1	0
V2:	1	0	-1

The first (V0) defines the overall level and the second two (V1, V2) are contrasts forming a complete comparison for the 'type' factor. Although we have chosen a version of a particular comparison in Section 2.15, contrasting scale 1 with each of scales 2 and 3 in turn, we shall see that for the purposes of this particular analysis it does not matter which way we choose to summarize the 'type' comparison. It is important only that we separate the overall level from the factor effects. Seen in this perspective, we may consider the previous analysis in Section A.2 of the groups' profiles across the three variables to be an analysis of the mean response level *within* each level of the 'type' factor. In the present analysis we are considering the differences *between* levels of the 'type' factor. To highlight the relationship of this analysis to that of Section A.2 we shall ask of the between-subjects design the same questions that we considered there, namely the tests of hypotheses about the Helmert contrasts between groups. We shall demonstrate the analysis in two stages, beginning with a simple multivariate analysis of one between-variables factor ('type') and one between-subjects factor ('experience'), although this first analysis does not precisely answer our primary question.

Table A.9 shows this analysis of variance. The between-variables factor produces two separate sets of new variables transformed from the original three scales. As stated above, these are

(a) the single variable representing the overall level of the 'type' factor (V0); and

(b) the pair of variables representing the factor effects – the differences between 'types' (V1, V2).

Note that in the between-subjects factor the degrees of freedom have been partitioned to provide two separate hypotheses, but in the between-variables factor there is no requirement, in this example, to partition the degrees of freedom. The effects of this factor are tested simultaneously as a complete comparison with two degrees of freedom.

Each set is analysed separately against the between-subjects design, exactly as in Section A.2. The first part of the table shows the tests for the first 'set', the single variable V0. This of course is therefore a simple univariate analysis, the sums of products matrix reducing to the familar sum of squares, and the analysis is performed in the usual way. The F-statistic of univariate analysis is equivalent

Alcohol relapse

Table A.9 Multivariate analysis of variance of the between-variables factor 'type' and the between-subjects factor 'experience' for the data shown in Table A.5 for partitioned between-groups degrees of freedom (showing the 'unique' contribution of each to the complete comparison).

(i) Analysis for the variable representing overall level of response (V0).

Source	d.f.	SS (V0)	F-statistic
Within groups	248	75 345.5	
Grand mean	1	339 561.0	
Between groups	2	2 745.1	4.518 (sig .012)
1st Helmert (1 vs. 2, 3)	1	2 729.9	8.985 (sig .003)
2nd Helmert (2 vs. 3)	1	165.0	0.543 (sig .462)

(ii) Analysis of the two variables representing the effects of the factor 'type' (V1, V2).

Source	d.f.	SP matrix (V1)	(V2)	Pillai statistic
Within groups (E matrix)	248	20 722 18 857	— 24 586	
Grand mean (H matrix)	1	18 399 30 112	— 49 280	.687 —
Between groups (H matrix)	2	277.57 307.78	— 342.07	.015 (sig .451)
1st Helmert (H matrix)	1	257.98 282.27	— 308.85	.013 (sig .188)
2nd Helmert (H matrix)	1	0.81 0.10	— 0.01	.0001 (sig .987)

(see Section 4.3) to the Pillai–Bartlett statistic (and in fact to all of the four multivariate statistics) when our multivariate 'set' consists of just one variable. Moreover, our definition of the original scales implies that the overall level V0, which is the total of the three scales, is identical to the RPI total score analysed in Section A.1. The reader may compare the analyses of Tables A.9(i) and A.3 to vindicate this statement, and we shall not comment further on these tests except to restate the conclusion that the higher overall level of perceived vulnerability distinguishes the first group from the others.

110

The second set of variables provides a bivariate analysis for the effects of the 'type' factor and this is shown in Table A.9(ii). It parallels exactly the format of the previous multivariate analysis in Section A.2 (Table A.7) except that the sums of products matrices now have only two rows and two columns, one for V1 and one for V2. Note that because the analysis deals with them as a bivariate pair, and not individually, the ensuing tests would produce the same results regardless of which particular contrasts were used to summarize the comparison of 'type' levels (Section 4.2 discusses this invariance of multivariate tests). The first entry after the within-groups error matrix is the hypothesis matrix for testing whether the grand mean across the whole sample is zero for both V1 and V2. This would simply be testing whether there are any effects on average due to 'type' – that is, whether there are differences between the three variables which constitute the levels of the factor. These are precisely the differences anticipated in the discussion above which will arise because the numbers of items in the original scales differ. We are not unduly concerned with testing this hypothesis, but note simply from the size of the test statistic that our expectations are met.

The next entry, which we will again ignore as we did in the previous analyses, relates to the test of the compound null hypothesis that no between-group differences exist (see Section 2.7) and we turn instead to the tests of each of the two separate hypotheses concerning the Helmert contrasts. Taking the ratio of each hypothesis matrix in turn to the error matrix produces the two test statistics shown. If we continue to use an individual test level of .01 type I error, we see that neither of these hypotheses is rejected. Thus we conclude, for each effect, that it is not required in addition to the presence of the other in order to explain the data. It is still possible that either effect, by itself, might be significant for we have only shown that the complete comparison is not required. However, we shall not proceed to these tests but instead shall pause to consider why it was stated above that this first-stage analysis does not answer our main question of interest, 'have we gained from using the three scales UM, ES and LV instead of one overall score?'

By testing in the manner above, for instance, the null hypothesis that the second Helmert contrast is zero on the responses, we are asking whether the mean vector for V1 and V2 in group 2 is equal to that in group 3. We are asking therefore whether the response variables V1 and V2, which themselves represent differences between the original scales, carry any information which will distinguish group 2 from group 3. Now the question in which we are interested clearly asks whether the differences between the original scales carry any *additional* information about the groups – information additional to that carried by the overall total score. We therefore do not want simply to test the 'type' effect, but need to test the 'type' effect when controlling for the overall level of response. (Sections 3.4 and 3.8 discuss this point.) It may seem strange to test for 'type' carrying additional information about an hypothesis when the previous analysis showed that it was anyway unable to provide enough

Alcohol relapse

Table A.10 Analysis of the two variables representing the effects of the factor 'type' (V1, V2) controlling for the variable representing the overall level of response (V0).

Source	d.f.	SP matrix		Pillai statistic
		(V1)	(V2)	
Within groups (E matrix)	247	14 450 8 189	— 6 583	
Regression (H matrix)	1	6 321 10 668	— 18 004	.819 (sig .000)
Grand mean (H matrix)	1	119.49 378.32	— 717.46	.163 —
Between groups (H matrix)	2	11.60 5.52	— 76.12	.037 —
1st Helmert (H matrix)	1	2 729.89 918.22	— 308.85	.035 (sig .013)
2nd Helmert (H matrix)	1	165.03 1.45	— 0.01	.011 (sig .252)

information to reject that hypothesis. However, such a situation is possible, since effects can be masked by the 'noise' of large experimental error and by other effects. We now proceed to a second analysis in which the 'type' main effect is tested controlling for the overall level by using it as a covariate, whereas the preceding analysis tested the main effect and the overall level completely separately.

Table A.10 shows this analysis and it is in the same form as the analysis of the 'type' effects in Table A.9(ii), except that an extra source of variation is included, that due to regression of the responses on the covariate. Thus we have a bivariate analysis of V1 and V2 ('type' differences) after allowing for the single covariate, V0 (overall response level). The entry for the regression relationship instead of being the sum of squares familiar in univariate regression, is a sums of products matrix which can be used as a hypothesis matrix in testing for a null regression.

A glance at this table shows from the error matrix that the within-group variation has been considerably reduced from that shown in Table A.9(ii). This reduction is due to the effect of the responses' regression on the covariate. This regression effect is tested in the usual multivariate manner, by using the matrix ratio $\mathbf{H \cdot E}^{-1}$ and the Pillai–Bartlett test statistic shows a very high level of significance. Passing to the tests for the individual Helmert contrasts, we see if each case that although the variation attributed to these sources has been reduced, the reduction in error variation has more than compensated. The test of the hypothesis that group 2 is equal to group 3 is not rejected, but the test that group 1 is different from their average is only fractionally short of achieving significance. We may conclude that even in a test at this high level of significance, there is evidence that the differences between the three types of response scales in the RPI provide extra information for distinguishing the first group of people (the 'long-standing' group) from the others.

The nature of this extra information is not hard to see. The profile analysis in Section A.2 showed the third scale, LV, to be the most important in dis-criminating between the 'long-standing' group and the others. It has, however, a very small range of scores compared with the other scales, so that a raw total of the three swamps its effect. Its differences from the other two scales are lost in their greater scale of error variation and these differences fail to discriminate between the groups. It is only when controlling for the highly variable level of overall endorsements that we can see the effects due to this minor scale. Thus it is important to separate the three types of perceived vulnerability in order to demonstrate the ability of the LV scale to distinguish the 'long-standing' relapsers from the rest.

Table A.11 Raw data for the variables used in the RPI.

	Grp	UM	ES	LV		Grp	UM	ES	LV
1	2	27	0	3	23	2	13	9	7
2	1	37	19	6	24	1	15	12	3
3	2	2	2	6	25	1	25	18	6
4	3	24	15	3	26	1	23	11	4
5	1	34	13	4	27	1	26	6	9
6	1	37	10	7	28	1	18	0	2
7	1	34	23	9	29	1	21	18	6
8	3	7	9	7	30	1	0	4	3
9	3	10	0	0	31	1	14	7	2
10	3	6	7	6	32	2	14	13	0
11	2	30	16	9	33	1	10	13	7
12	3	24	21	8	34	1	13	18	9
13	1	22	6	8	35	3	41	16	9
14	1	24	7	7	36	1	24	15	3
15	2	12	14	4	37	2	1	6	3
16	1	31	13	9	38	2	15	16	7
17	2	2	3	3	39	2	6	2	0
18	3	3	4	0	40	1	34	16	6
19	1	26	9	9	41	1	37	19	9
20	1	33	14	9	42	2	2	4	0
21	1	18	7	6	43	1	28	6	9
22	1	16	8	9	44	2	21	15	8

	Grp	UM	ES	LV		Grp	UM	ES	LV
45	3	13	3	3	93	1	24	9	8
46	1	32	9	8	94	1	37	19	9
47	2	25	2	9	95	2	3	2	0
48	3	0	0	0	96	1	16	12	8
49	2	27	13	4	97	2	14	5	7
50	1	12	10	4	98	1	35	16	6
51	2	11	11	7	99	1	35	15	7
52	1	30	13	6	100	1	34	21	9
53	2	32	19	5	101	1	34	23	7
54	1	9	7	1	102	2	31	7	9
55	1	32	24	9	103	1	35	10	5
56	2	31	18	5	104	1	24	7	6
57	3	26	16	3	105	1	29	16	9
58	3	0	2	1	106	1	17	3	9
59	1	5	3	4	107	1	11	12	6
60	1	15	5	3	108	2	39	23	9
61	1	29	18	4	109	1	21	10	9
62	1	28	21	9	110	1	11	3	4
63	2	16	14	9	111	2	33	17	9
64	3	15	5	8	112	2	0	0	0
65	1	27	15	7	113	2	31	14	7
66	2	21	12	5	114	1	18	14	2
67	1	42	23	9	115	3	41	12	9
68	2	21	12	9	116	1	5	10	9
69	1	16	0	3	117	3	30	12	6
70	1	26	0	9	118	2	3	9	6
71	1	20	2	5	119	2	9	8	2
72	1	1	1	1	120	1	17	20	7
73	3	15	23	9	121	3	0	10	3
74	1	16	13	3	122	2	22	10	5
75	2	6	6	4	123	3	20	6	1
76	1	19	13	6	124	3	14	3	4
77	1	34	23	9	125	1	7	7	3
78	1	26	21	7	126	3	14	12	5
79	2	28	22	7	127	2	14	1	5
80	2	36	21	9	128	2	40	5	5
81	1	19	16	9	129	1	24	9	7
82	1	32	18	8	130	2	15	5	5
83	2	3	5	4	131	1	8	17	6
84	1	9	16	9	132	2	5	4	3
85	2	0	6	7	133	1	25	1	9
86	1	8	4	5	134	2	17	2	9
87	1	42	22	9	135	1	17	18	8
88	1	41	21	8	136	1	26	18	8
89	2	16	10	2	137	1	29	10	6
90	2	7	12	4	138	1	26	15	9
91	2	36	11	7	139	1	999	999	999
92	3	18	10	9	140	2	20	14	9

	Grp	UM	ES	LV		Grp	UM	ES	LV
141	1	10	17	9	189	1	35	15	9
142	3	31	20	6	190	2	17	5	5
143	1	18	7	3	191	2	29	16	7
144	1	22	13	4	192	2	32	20	8
145	2	21	9	6	193	1	13	5	3
146	2	3	7	4	194	2	37	22	9
147	2	16	1	0	195	1	26	13	6
148	1	15	12	7	196	1	17	9	4
149	3	7	5	1	197	1	28	10	5
150	2	18	12	4	198	3	28	8	6
151	1	27	11	9	199	3	5	3	0
152	1	18	9	8	200	2	25	19	8
153	2	0	0	0	201	3	19	23	3
154	1	8	14	6	202	2	33	14	5
155	2	7	7	6	203	2	7	17	4
156	1	7	4	5	204	1	19	13	6
157	2	19	21	8	205	1	32	13	9
158	2	30	16	8	206	3	0	3	0
159	3	6	16	8	207	1	35	18	999
160	1	28	13	9	208	2	16	18	2
161	1	17	8	3	209	1	39	18	9
162	2	17	13	6	210	1	28	20	3
163	1	4	4	3	211	2	40	4	3
164	2	12	16	6	212	1	18	9	8
165	1	2	1	3	213	2	11	11	6
166	2	15	3	6	214	1	11	21	9
167	1	28	21	6	215	2	38	999	8
168	1	14	10	2	216	2	23	12	7
169	2	13	14	7	217	1	32	20	9
170	2	0	0	0	218	2	14	3	3
171	2	0	6	2	219	2	10	6	4
172	1	6	1	2	220	3	32	7	4
173	1	35	16	9	221	1	999	999	999
174	1	23	3	8	222	2	22	19	5
175	3	10	5	6	223	1	0	0	0
176	1	29	20	9	224	2	29	16	6
177	1	23	12	6	225	1	0	10	8
178	2	4	0	2	226	1	23	9	7
179	1	11	13	8	227	1	37	22	9
180	2	34	16	7	228	2	25	17	6
181	3	0	0	0	229	1	17	10	3
182	1	26	5	8	230	2	18	3	2
183	3	15	15	5	231	2	19	22	8
184	1	3	5	2	232	1	16	4	5
185	3	34	11	7	233	2	0	2	2
186	2	6	19	6	234	1	24	16	8
187	1	18	12	5	235	2	24	14	6
188	1	3	5	7	236	2	19	5	7

Table A.11 (continued)

	Grp	UM,	ES	LV		Grp	UM	ES	LV
237	3	33	19	3	247	2	22	20	7
238	1	15	10	7	248	1	21	15	6
239	1	24	16	7	249	2	17	9	6
240	2	33	14	7	250	1	28	23	9
241	1	26	3	5	251	1	15	17	9
242	2	26	12	8	252	3	11	5	6
243	1	26	16	6	253	3	39	15	6
244	2	40	20	9	254	2	30	17	9
245	3	33	17	7	255	2	22	11	3
246	2	30	13	8					

Salsolinol ——————— Study B
excretion

B.1

This example is taken from a study of the effects of drinking alcohol, carried out by Dr A. Topham (Addiction Research Unit, London, and Addenbrooke's Hospital, Cambridge) who has kindly allowed us to use her data. In cases of excessive indulgence in alcoholic drink a bodily dependence will be set up which appears to cause the formation of various alkaloids having a similar structure to heroin and morphine. One such is salsolinol, which is suspected of interacting with neurological agents in the body. It is this biochemical agent that the research project was designed to study.

To study the role salsolinol plays in bodily dependence on alcohol, 16 individuals attending an alcoholism treatment unit were observed over a period of four days immediately after being admitted to the unit. Two of the 16 failed to complete the study, for reasons that were adjudged to be unrelated to the enquiry; these individuals were dropped from the subsequent analysis. Of the 14 individuals who were observed for the four days, eight of these were considered to be 'severely' dependent, and the remaining 6 only 'moderately' so, this categorization being a clinical judgement made on the basis of their symptomatology. All had been drinking alcohol heavily prior to the study period but throughout the study period in the treatment unit they drank no alcohol. The questions of central interest were: would there be any systematic changes in their salsolinol levels over the study period; and would the 'severely' dependent group differ from the remainder, in having a generally higher level of salsolinol? The measurement of salsolinol was made from a biochemical analysis of urine samples, taken daily throughout the study period.

Thus the analysis consists of a between-groups design of just one factor, dependence', at two levels, 'severe' and 'moderate'. The response variables are the four repeated measures representing salsolinol excretion (one for each day of the study).

An initial inspection of graphical plots of the distributions showed a heavy skew in all of them. A logarithmic transformation of these raw measurements was enough to make them approximately normally distributed, at least as far as could be judged by eye. The small number of observations involved in the analysis was thought to render any intensive study of the normality of the log-transformed data a wasted effort. The means of these data are shown in

Salsolinol excretion

Table B.1. Note that the effect of reducing the skew by the transformation is to produce a simpler pattern of daily mean values.

The analysis of variance can be carried out by univariate methods (see Section 3.6) but it will be treated here as a multivariate analysis with four response variables, one for each day's salsolinol excretion rate. To study the pattern of the time effects in the response it would be inadequate merely to carry out four univariate analyses to show day-by-day differences. Similarly, it would be inadequate to ask the general question 'do the groups differ on the responses?'; such an analysis would not allow an easy characterization of any patterns of

Table B.1 Mean levels of daily excretion of salsolinol for two groups of alcohol-dependent individuals. (The full data are given in Table B.5.)

	Day 1	Day 2	Day 3	Day 4
Raw data (mmol)				
Group 1 (moderate)	1.635	1.842	1.753	1.980
Group 2 (severe)	1.909	1.569	2.762	3.737
Log-transformed data				
Group 1 (moderate)	−0.110	0.103	0.356	0.441
Group 2 (severe)	0.116	0.244	0.723	1.036

differences that might be found. Instead, the analysis is treated as a structured set of questions (see Section 3.5) about the relationships between the variables. Thus the four days' measurements are considered as four levels of a single factor, 'time', rather than as four separate variables.

To characterize the patterns of change over the 'time' factor any predetermined combinations of the set of variables may be used including, in particular, any of the summary sets of contrasts for a comparison of factor levels given in Section 2.15. To answer the questions central to the present study, an obviously suitable choice for representing the complete comparison (see Section 2.3) of the 'time' factor is the set of orthogonal polynomial trends (Section 2.15, para. 5). This set, describing differences between levels, will need to be augmented by the addition of the overall mean level of response in order to give a full non-singular transformation of the original four variables into an equivalent set of new variables (see Section 3.7). The resulting coefficients in the transformation for the creation of the new trend variables for the 'time' factor, which shall be called M, L, Q and C, can be written out as a matrix, as in Section 2.15, namely:

M: 1   1   1   1
L: − 3   − 1   1   3
Q: 1   − 1   − 1   1
C: − 1   3   − 3   1

118

The analysis, then, is a structured multivariate analysis of variance with one between-variables factor, 'time', at four levels. It will use a predetermined transformation of the variables (an orthogonal polynomial transformation) to represent the complete comparison of the factor effects, giving four new variables: one representing the mean level over the four days, and the other three representing the linear, quadratic and cubic trends in the response profile. The between-subjects design is one factor, 'dependence', at two levels, moderate and severe, indicating the two groups of individuals. This factor therefore has a single degree of freedom and therefore a single contrast representing its effect:

$$1 - 1$$

The analysis of variance table is given in Table B.2 for these polynomial trends. The first variable, mean level (M), represents the average score of each

Table B.2 Multivariate analysis of variance of the repeated measures factor 'time' and the between-subjects factor 'dependence' in salsolinol data.

(i) Analysis for the variable representing overall level of response (M).

Source	d.f.	SS (M)	F-statistic
Within groups	12	29.26	
Grand mean	1	33.61	
Between groups	1	6.05	2.483 (sig .141)

(ii) Analysis for the three variables representing the effects of the factor 'time' (the polynomial trends L, Q, C).

Source	d.f.	SP matrix (L)	(Q)	(C)	Univariate F-statistic
Within groups (E matrix)	12	330.50	—	—	—
		−20.08	16.56	—	
		99.84	−12.07	208.63	
Grand mean (H matrix)	1	99.70	—	—	3.620 (sig .081)
		1.92	0.04	—	0.027 (sig .873)
		−14.30	−0.27	2.05	0.118 (sig .737)
Between groups (H matrix)	1	6.11	--	—	0.222 (sig .646)
		1.43	0.33	—	0.241 (sig .632)
		−1.40	−0.33	0.32	0.018 (sig .894)

subject over the 'time' factor (i.e. over the four days). It is customary to present these mean level results first, as a separate analysis, showing the analysis of variance between subjects' overall levels. The remaining three variables give the linear (L), quadratic (Q) and cubic (C) response profile components of each subject's pattern of day-to-day variation; these variables are dealt with subsequently as a set showing changes over 'time'. With so few observations in the data set, it was considered that significance tests should be conducted at the 10% level, instead of the more familiar 5% level. Although the type I error is greater, the accompanying type II error will be brought down from an otherwise overly high level.

The first part of Table B.2, then, shows the analysis for the variable M; it is therefore the usual univariate analysis of variance table. The error term for this analysis is the subjects' within-groups sum of squares, as in any one-way analysis of variance, and is the first entry. The second entry is the sum of squares for the grand mean of the data. The test of the null hypothesis that this is zero is not of interest in the study, and no F-value is shown for this test. The third entry is for the between-group differences, the main effect of the 'dependence' factor. The F-value for these group differences achieves only a 14% probability level and is thus not significant in a 10% significance test.

However, in this particular study it was felt that enough theoretical knowledge existed to justify a one-sided alternative hypothesis – that is, it is certain that group 2, the severely dependent individuals, *would not have* a lower excretion level than group 1, the moderately dependent individuals, but *might have* a higher level. Thus it could be argued that the alternative hypothesis of interest in this study is the one-sided hypothesis (that group 2 has a greater mean level than group 1) and not the usual two-sided alternative (that group 2 has a mean level different from group 1). If this a priori assumption is used, the 14% probability level on the F-test (which tests against a two-sided alternative hypothesis) can be considered as corresponding to a one-tailed t-test (against the one-sided alternative hypothesis) with a 7% level of significance.

In the second part of Table B.2 the three trend variables are presented to show the analysis of variance of the time profiles within subjects. The sources of variation shown in this part of the table are the same as those given in part (i) and as the first entry we again have the subjects' within-groups variation. This is the relevant source of error variation in the analysis and the appropriate matrix of sums of products relating to the three variables L, Q, C is shown. As described in Section 3.1, the diagonal of this matrix is composed of the three sums of squares that would be used in separate univariate analyses of each of the variables.

The next entry in the table is for the grand mean of the variables L, Q, C and should not, strictly speaking, be considered until after the bottom entry has been inspected. This last entry is used in testing for differences between the groups of subjects and the existence of these may well make irrelevant the mean values across all subjects. However, for the purposes of exposition it will be more easily dealt with at this point. The second entry, then, is the hypothesis matrix which relates to the existence of overall trends – that is, whether the linear trend (L) has a non-zero mean across all subjects, and similarly for the quadratic (Q) and cubic (C) trends.

Clearly a test of the overall null hypothesis 'all the average trends are zero' (and hence all 'time' differences, too, from which they are derived) could be made against the error matrix by a multivariate test (on the three trend variables). However, when dealing with questions about a set of increasingly complex trends such as the polynomials that are being used in this example, it is not likely that the researcher's interest is simply in whether any of them exist. Rather, it is a question of how complex the explanation of the data needs to get: a sequence of tests is called for and not a simultaneous test of all trends (see Section 5.4).

Thus, rather than needing to perform a fully multivariate test of hypothesis, the problem in fact concerns a 'multiply univariate' set of sequenced questions (see Section 3.5). In a multivariate test the whole matrix (including the off-diagonal terms representing the relationships between the variables) would be used simultaneously. In these univariate tests only the values in the diagonal of the matrix will be used, to produce a series of univariate F-tests. Thus the original structured multivariate problem of describing the pattern of daily differences has been reduced, by virtue of the polynomial trend transformation, to a sequence of univariate questions.

The sequence of univariate F-tests is therefore used to test each trend in turn, aiming to provide a more useful analysis than the overall multivariate procedure. It is these F-test statistics which are shown in Table B.2. Beginning at the foot of the matrix, and testing each time at the 10% level, it can be seen that the cubic trend does not contribute significantly to the time profile. The quadratic in its turn also is not significant, but a significance is attributed to the linear trend (with a probability $<.081$). Note that immediately any of these tests produces a significant result the sequence is interrupted, and no further trends tested. Thus if the quadratic had been a significant trend the time profile would have been declared to be 'of quadratic order', and the linear trend not tested. In fact, the significant result for the linear trend is the only evidence that there is any consistent pattern across time in the responses.

Note that the overall type I error for a sequence of tests such as this is greater than the type I error used at each step (see Section 5.4). The formula given there shows the overall error for three independent tests each carried out at the 0.10 level is

$$1.0 - (1.0 - 0.10)^3 = 0.271$$

In these trend tests this level provides an upper bound to the overall significance level, since the tests are not in fact independent.

Salsolinol excretion

Turning to the final entry in Table B.2(ii), which was left in abeyance above, a similar procedure can be carried out in relation to group differences in trends. This hypothesis matrix is used to test whether the groups differ on the trends defined by variables L, Q and C – that is, whether 'dependence' affects the 'time' effects. (The immediately preceding paragraphs dealt with the values of these 'time' trends averaged across both dependence groups.) The relevant hypothesis matrix is treated in exactly the same way as the hypothesis matrix for the average trends across groups above. In fact, the diagonal of the matrix consists of values too small to yield significance at any reasonable test level and we shall not detail the ensuing tests, which are all non-significant. The conclusion is that there is no evidence of a difference between the two 'dependence' groups (apart from the significant difference in the overall response level, M, previously identified).

The estimated values of the grand means of the variables L, Q and C – that is, the trends in the 'time' effects – are, respectively:

0.575   0.015 -0.081

They are not usually directly interpretable, but can be used to reconstruct a simplified, fitted trend to the observed means. The between-group difference on variable M – the effect of 'dependence' on the overall level of response – is estimated as -0.332. Hence the simple explanation of the data to emerge from this analysis is that there is possibly a linear trend common to both groups showing increasing salsolinol excretion over the four days; but that the 'severely dependent' group have this trend superimposed on a generally higher overall level than the 'moderately dependent' group. The apparently steeper rate of increase in the 'severely dependent' group that can be seen from inspecting the table of means (Table B.1) is not large enough to be significant of a genuine population difference in rates of increase.

Note that with a study having as few subjects as this, it can be preferable to carry out statistical tests at a very lax type I error level, such as 20% or even 25%. The effect of this strategy is to improve the power of the tests to detect reasonable values of the means under the alternative hypothesis. A stricter 5% testing strategy, for instance, has very little chance of detecting anything but the grossest departures from the null hypothesis, considering the sample size and the size of the error variation. This improvement should be carefully balanced against the weaker confidence attaching to statements made under the null hypothesis and the associated high type I error rate. Certainly in pilot studies and other exploratory work such a strategy can be most useful.

B.2

As an alternative strategy, the researcher could attempt to 'buy' precision at the expense of making further assumptions: if the data warrant the requisite independence assumptions about the distributions of the responses, a univariate approach to repeated measures analysis could be employed (see Section 3.6). It is worth underlining at this point the relationship between the analysis in Section B.1 and the 'standard univariate method' of analysing the same repeated measures problem. To do this it is necessary to re-present the analysis shown in Table B.2 with one significant change.

The 'time' comparison was characterized in the preceding analysis as the polynomial trends, for which the transformation matrix was shown. Now a small complication is introduced into that representation: normalized contrasts will be used (see Section 2.2) instead of the ones shown above. Recall that rescaling all the weights in a contrast has no consequences for the comparison it is used to define, except to rescale the estimated effect. The hypothesis and error sums of products are altered equally by the rescaling, so that significance tests are unchanged. This can be seen in Table B.3, which shows the analysis equivalent to that in Table B.2, but for variables M, L, Q, C defined by the following transformation:

M:	.500	.500	.500	.500
L:	$-.671$	$-.224$	.224	.671
Q:	.500	$-.500$	$-.500$	.500
C:	$-.224$	.671	$-.671$	.224

These are the normalized contrasts obtained from the orthogonal polynomials given in Section 2.15, para. 5, and the reader may check that the sum of squares for each row is unity. Normalizing is usually a standard option in a computer program for analysis of variance. The point of making a normalized transformation is that, whereas significance tests are invariant under all linear combinations of variables (Section 4.2), orthonormalized transformations have the further property of leaving unchanged the original total sum of squares of the data *taken across all variables*. The transformation of the preceding analysis in Section B.1 did not preserve this sum of squares. (This total can be calculated from summing all the diagonal terms of all the matrices in Table B.3.)

Inspecting Table B.4 will make clear the relevance of this point. This table shows a standard univariate analysis of variance for these data. The entries under the 'source' heading are simply a relabelling of those in Table B.3, with the entries for error appearing last. The sums of squares for each section of the table (overall mean, groups, time, etc.) are the sums of the diagonal terms of the relevant matrices; these diagonal elements themselves are the entries for the partitioning into linear, quadratic and cubic components. Note that the degrees of freedom in this table are much greater than those of Table B.3. There is now a total of 56 degrees of freedom in the whole table, four times the total number in Table B.3.

The F-tests for the factors and interactions compare these totalled sums of squares (and degrees of freedom) with the total error. They are often obtainable directly from a multivariate analysis of variance program under a name such as

123

Salsolinol excretion

Table B.3 Re-analysis of the multivariate analysis shown in Table B.2, using ortho-normalized contrasts for the comparison for 'time'.

(i) Analysis for the variable representing overall level of response (M).

Source	d.f.	SS (M)	F-statistic
Within groups	12	7.314	
Grand mean	1	8.402	—
Between groups	1	1.514	2.483 (sig .141)

(ii) Analysis for the three variables representing the effects of the factor 'time' (ortho-normalized polynominal trends L, Q, C).

Source	d.f.	SP matrix (L)	(Q)	(C)	Univariate F-statistic
Within groups (**E** matrix)	12	16.525	—	—	—
		−2.245	4.140	—	
		4.992	−1.350	10.432	
Grand mean (**H** matrix)	1	4.985	—	—	3.620 (sig .081)
		0.215	0.009	—	0.027 (sig .873)
		−0.715	−0.031	0.103	0.118 (sig .737)
Between groups (**H** matrix)	1	0.306	—	—	0.222 (sig .646)
		0.160	0.083	—	0.241 (sig .632)
		−0.070	−0.037	0.016	0.018 (sig .894)

'Averaged F-tests'. It can be shown that ignoring the off-diagonal terms of the trivariate matrix for L, Q and C, as such summated tests do, gives a valid test only if the variables concerned are uncorrelated and have equal variance (see Section 3.6) – these are the assumptions with which the extra degrees of freedom are 'bought'. Note that, unlike Table B.3, even when testing the partitioned effects of L, Q and C for each of the sources of variation in Table B.4, the F-test is usually made against the *pooled* error term (subjects-within-groups by time) with 36 error degrees of freedom. These extra degrees of freedom are obtained by making the same further assumptions. In this study, a simple calculation on the error matrix in Table B.3(ii) shows one correlation as high as .72 and a ratio of the largest to smallest variances of 1.959. Although neither a test of sphericity nor one using the Fmax criterion reject these assumptions, in view of their large-sample validity and the few degrees of freedom available in this study it is a moot point as to which analysis is to be preferred.

124

Table B.4 Repeated measures analysis of the above data by the univariate method.

Source	d.f.	SS		F-statistic
(i) *Between subjects* (1 variable: mean level)				
Overall mean	1	8.402		13.784 (sig .003)
Groups	1	1.514		2.483 (sig .141)
Subjects within grp	12	7.314		—
(ii) *Within subjects* (time trends)				
Time	3	5.097		1.967 (sig .136)
(L)	1		4.985	
(Q)	1		0.009	
(C)	1		0.103	
Groups × time	3	0.405		0.156 (sig .925)
Grps × (L)	1		0.306	
Grps × (Q)	1		0.083	
Grps × (C)	1		0.016	
Sub within grps × time	36	31.096		—
Within grp × (L)	12		16.525	
Within grp × (Q)	12		4.140	
Within grp × (C)	12		10.432	

Table B.5 Raw salsolinol data and dependence status for the 14 individuals in the study (in mmol).

Grp		Day 1	Day 2	Day 3	Day 4
1	2	.64	.70	1.00	1.40
2	1	.33	.70	2.33	3.20
3	2	.73	1.85	3.60	2.60
4	2	.70	4.20	7.30	5.40
5	2	.40	1.60	1.40	7.10
6	2	2.60	1.30	.70	.70
7	2	7.80	1.20	2.60	1.80
8	1	5.30	.90	1.80	.70
9	1	2.50	2.10	1.12	1.01
10	2	1.90	1.30	4.40	2.80
11	1	.98	.32	3.91	.66
12	1	.39	.69	.73	2.45
13	1	.31	6.34	.63	3.86
14	2	.50	.40	1.10	8.10

Slimming —————————— Study C
clinics

C.1

Simple overeating is a central cause of general obesity and there exist various groups or clubs to help people control their weight. These groups aim to encourage people to control their diet by offering general encouragement and support through regular meetings, discussion groups, and other such activities, including the showing of relevant films or the provision of lectures by a professional dietitian or health expert. The project described here was devised by a group of behavioural psychologists in conjunction with health visitors to see if the effectiveness of the package offered by such groups could be improved. We are grateful to Professor R. Hodgson and H. Rankin for permission to use these data, and to J. Stapleton who did the preliminary analyses.

The aim was a simple one. A technical manual would be added to the overall package which would offer advice based on psychological behaviourist theory and which, it was hoped, would help the would-be weight losers in their attempts to control their eating. The content of the manual was drawn up after an examination of the psychological literature, which suggested that self-monitoring, introducing personal coping strategies, and reinforcing any small successes were all effective ways of changing eating habits. The manual also included a chapter of advice on how to prevent putting the lost weight back on again.

Two different health clinics were selected which ran such courses lasting two months with one attendance per week during that period. Both clinics ran two such courses concurrently at different times during the same evening of the week. It was decided, therefore, to use the manual as an addition to the regular package offered by these clinics in one of the groups from each clinic and to leave the other group as a control group for comparison purposes. Information was collected on the clients of the clinic, firstly prior to the commencement of the course; secondly, nine weeks later at the end of the course; and thirdly, three months after the commencement of the course. It was also planned to trace the participants in the trial one year later in order to establish the long-term effectiveness of the project.

Neither of the two clinics had a set diet or slimming plan and subjects were left to choose for themselves the way in which they proceeded. The manual was considered as a blanket addition to the package offered by the clinic and no

126

attempt was made to change people's strategies directly, nor was any attempt made to ensure regular attendance over and above the clinics' usual procedures. However, checking was carried out each week to ensure that the group receiving the technical manual had actually read the relevant sections of the manual. The general discussion sessions also included opportunities to discuss points or queries that arose from the technical manual. The trial therefore aimed at two clinics and two conditions, experimental and control, in each of these clinics.

It was thought important to consider further the selection within each of the courses of participants who would be suitable for the trial. Firstly, it was necessary to exclude individuals who would not usually be considered over-weight. Each individual's weight was transformed into a percentage overweight figure by reference to standard health tables, and those who were overweight were included in the trial. There was also concern that the attenders at the slimming clinic were, in some sense, 'experienced' slimmers. A further set of individuals was therefore recruited from the waiting-lists of the clinics to provide 'novice' slimmers. The precise criteria for these two categories were:

- an experienced slimmer was one who had been trying to slim for more than one year;
- a novice slimmer was one who had been trying for no more than three weeks.

All the volunteers participating in the trial were female bar one, who was excluded since clearly too little information was available to analyse the differences between sexes reliably. (Those who were excluded from the trial still participated in the course as usual, but simply were not included in the analysis of the results.)

Participation in the trial was entirely voluntary, but the number of people who met the criteria and refused to co-operate was minimal. A small number of people refusing to participate is unlikely to introduce an appreciable bias into the data, but in general we must assume such refusals occur randomly. A strong systematic effect could result if this self-selection were associated with characteristics that influence the response variables. It would imply that in effect the population under study was not simply 'slimmers', but rather those with the particular characteristics that lead to self-selected participation in experimental studies.

At each occasion the percentage overweight of the individual was calculated, along with other measurements of obesity – namely the girth of bust, waist, hips, thigh and arm. When interest focuses on changes in physical measurements of this sort it is often the case that percentage reduction or increase is a more appropriate measure than an absolute reduction or increase. A large person has, so to speak, more to lose than a small person. For these data, it was arbitrarily decided a priori to use the baseline measurements taken prior to commencement of the course as an initial position from which all subsequent measurements were calculated as a percentage drop (or increase).

Slimming clinics

Table C.1 gives some support to this statement. It shows the distribution of subjects for percentage overweight initially, at the commencement of the course, and subsequently at the three-month follow-up point. Both distributions show a strong skew. However, the third distribution is of percentage improvement; that is, follow-up overweight divided by initial overweight (multiplied by 100), and is seen to be fairly symmetric. We should note that this distribution is not required to be normal, since it is a mixture of all the observations from groups with mean values that are probably different. It is the error distribution in the model which should be a normal distribution, and the overall distribution will be dispersed across a wider range because of the spread between the means.

Table C.1 Distributions for 'overweight' measures.

Distribution of initial percentage overweight.

Mid-point	105%	115%	125%	135%	145%	155%	165%	175%	185%	195%
Initial frequency	16	15	22	12	6	3	4	1	0	1

Distribution of 3-month follow-up percentage overweight.

Mid-point	105%	115%	125%	135%	145%	155%	165%	175%	185%	195%
Follow-up frequency	21	19	15	6	3	1	1	1	0	1

Distribution of percentage improvement (follow-up/initial percentages).

Mid-point	89%	95%	91%	97%	103%	109%	115%
Frequency	1	4	12	32	16	2	1

The final set of response variables used for the trial is therefore firstly improvement in percentage overweight, measured at nine weeks, three months and a long-term follow-up at one year; and secondly, five variables measuring percentage improvement in girth, taken at two points in time (nine weeks and three months). The between-subjects factors (see Section 2.2) are:

(1) clinic: A and B;
(2) condition: experimental or control; and
(3) status: experienced or novice.

The full data are given in Table C.18. In the researchers' first analyses the relationship between the two follow-up points (of nine weeks and three months) was not of particular importance, the first being considered as an immediate post-course measurement of short-term engagement in the scheme and the second a medium-term outcome of the ability to maintain the techniques learned over the course itself. There was no reason for combining the information from these two different time points into a single analysis, since the researchers' main questions were confined to medium- and long-term outcomes. The presentation here is of only the three-month follow-up data (the other set follows an identical structure). Testing the experimental versus control conditions is initially performed on the overweight improvement variable only, and the analysis is presented in Tables C.2, C.3 and C.4. In Section C.2 these tests are paralleled by tests on the set of five girth improvement measures. A third analysis (Section C.3) brings together these two sets of data; and the long-term follow-up trends are analysed in Section C.4.

The first analysis, then, is a univariate analysis of variance for a three-factor between-subjects design (see Section 2.11). Table C.2 gives the simple means for overweight improvement in each cell of the condition by status by clinic cross-classification. Since each factor has only two levels, the only contrast needed for representing the complete comparison of its levels (see Section 2.2) is, in all three cases,

1/2 $-1/2$

Contrasts representing the interactions between each pair of factors and the three-factor interaction will be constructed from this basic contrast, using the computer program (as described in Section 2.11). The result of using all these effects and interactions in the analysis is to saturate the model – that is, they form

Table C.2 Overweight percentage improvement at 3 months: mean values (and numbers) in the cells of the clinic by status by condition cross-classification.

	Status: Experienced		Status: Novice	
Condition	Clinic A	Clinic B	Clinic A	Clinic B
Experimental	92.7 (5)	98.2 (5)	99.1 (12)	99.7 (9)
Control	93.4 (11)	96.2 (4)	99.2 (6)	97.2 (13)

Table C.3 Analysis of variance for the data in Table C.1.

Source	d.f.	SS	F-statistic	Significance
Within cells	58	1 740.48	—	
Constant	1	1 437.56	11.21	.001
Clinic	1	44.08	1.44	.235
Status	1	212.67	6.96	.011
Condition	1	35.08	1.15	.288
2-factor interactions	3	100.65	1.09	.433
3-factor interaction	1	.004	.004	.995

Slimming clinics

a 'complete comparison' for the composite 8-cell design. Under these circumstances, the estimates of the effects are provided by applying the coefficients of the composite contrasts directly to the cell means; it should be noted that the main effects estimated in this way are estimated *allowing for interactions* and as such are often either not the most useful or the most efficient.

Table C.3 shows the between-subjects design and gives the standard univariate break-down of the variation in the data. The question of central interest is whether the 'condition' factor had any effect – do treatment and control groups differ? The other two factors, 'clinic' and 'status', were introduced initially to increase the size of the database; now they are retained in the analysis because they might be associated with different levels of responses and introduce distortion into the central comparison between treatment and control. Any effects they might have would also cause increased variation in the data, which if ignored would inflate the within-group error sum of squares and hence make the experiment less sensitive.

It was not anticipated that any strong interactions would occur, but these were none the less checked in case expectations were wrong. A 5% significance level was chosen for the tests. The three-factor interaction was tested first, being the highest-order effect in the design. The sums of squares shown in Table C.3 have been computed sequentially (by the 'sequential method', as some programs call it; see Section 2.13). Each shows the effect of its associated comparison when controlling for the effects of comparisons higher up the table. The first null hypothesis to be tested is that the three-factor interaction effects are zero, and the sum of squares when compared with the within-cell sum of squares (or rather the mean squares) gives a non-significant value for the *F*-statistic. Thus we conclude that the three-factor interaction (controlling for main effects and two-factor interactions) is zero, and not necessary to explain the observed data.

Next, the three two-factor interactions were all pooled into a single test with (the pooled) three degrees of freedom, since no particular interest was attached to individual interactions. Again the interactions are not significantly different from zero, and thus the apparently large interaction effects, particularly between status and clinic, in the table of means (Table C.2) are in fact attributable to chance. We conclude that an adequate model could be produced from simply the main effects (see Section 2.10).

Continuing the sequence, the test of the comparison of central interest – the effect of the 'condition' factor – is now made. It is tested controlling for the effects of the other two factors, since we are interested in its role as an explanatory factor in the observations. Comparing its associated mean square with the error mean square shows it to have no effect on weight loss. We note in passing that the next line above in Table C.3, for the 'status' comparison, shows that the main effect of 'status' controlling for 'clinic' differences (but not controlling for 'condition') does achieve significance. The comparison for clinics indicated by the next entry in the table is the simple effect, not controlling for any of the other factors. It is not significant, but note that we are not interested in this test – it does not control for 'status', which has been shown to produce a significant effect.

Which other factors need be controlled for when testing the 'clinic' comparison is a moot point, for 'condition' shows no effect on the response and 'status' does show a significant effect. Generally the decision should depend on what a priori hypotheses were of interest. Given the researchers' initial expectations in this study, an analysis was performed in which the comparison for each factor's main effect was computed controlling for both other main effects. Table C.4 shows the analysis of variance for the three main effects, in which each sum of squares shown is this 'unique' attributable sum of squares for the factor comparison. Note that as far as the test of 'condition' is concerned, this *is* what was tested in Table C.3. In fact, no appreciable

Table C.4 Analysis of variance for the revised data from Table C.1 (unique sums of squares computations).

Source	d.f.	SS	F-statistic	Significance
Within cells	57	1 740.48	—	—
Constant	1	637.32	20.87	.000
Clinic	1	18.03	0.59	.445
Status	1	186.73	6.12	.016
Condition	1	35.08	1.15	.288

differences from Table C.3 appear in the sums of squares for the other main effects. We conclude that 'status' does affect the response, but that the other two factors do not.

As a final passing comment we should note the entry for the constant, or grand mean, in Table C.3. The sums of squares and the very highly significant F-test based on it is a test for whether the overall level of the response is zero. Since we expect it to be around, or somewhat below, 100%, this is not a useful or surprising result. If, however, we were to subtract 100 from our percentage overweight improvement, we would have a new response variable for which a zero value meant 'initial measurement is equal to follow-up measurement'. The overall level of this new response variable would indicate whether, on average, there was any change in weight. A test of whether this overall level is zero is clearly more interesting; it asks if we have observed any weight loss at all, on average, among all the subjects in the experiment. The repeat analysis shown in Table C.4 was in fact made on this new response variable, and it shows a significant loss in weight overall in the experiment. Note that none of the other sums of squares in Table C.4 will change as a result of subtracting 100 from the response variable, since they deal with *differences* between groups and do not depend on the absolute response level.

The implications of the constant in the between-subjects design being significantly non-zero needs to be considered with some care. Because there is a significant difference between the two types of slimmers, novice and experienced, the overall grand mean of the sample may not be easily interpretable. In particular, if the two groups are in some sense representative of a population in their relative sizes, the average is representative of the average in that population – a simple descriptive effect (see Section 2.9). In this study the two groups' relative sizes were manipulated by the experimenter and such an interpretation is not available.

Thus in spite of the overall weight loss, no differential effects on weight loss were found between experimental and control groups in this study, but differences between the two 'status' levels of novice and experienced were revealed. It remains to present the actual estimate of the effect of status on the result. The estimate associated with the model given in Table C.4 can be an inefficient one, since in that analysis allowance was made for the distorting effects of the other two factors, 'condition' and 'clinic'. If we decide, on the basis of the lack of evidence shown in the significance tests, that no such effects exist, then a better estimate of the difference is found by using the smallest model that has been found adequate to explain the data. In this case it is the analysis that includes only the overall constant and the effect for 'status'. Note that this analysis is in effect using a simple one-factor model with two levels to the factor. The estimated effect will therefore simply be one-half of the difference between the average scores of novice and experienced subjects, that is, the value of the contrast

$$1/2 \qquad -1/2$$

applied to the two groups when pooled over the other factors. Since the effects of the two factors we are omitting are small enough to be non-significant, we would not expect a large difference between these two estimates of the 'status' effect. In fact, when calculated they were -1.79 and -1.98 respectively.

C.2

We have used the weight loss of the individuals in this study of slimming clinics as an overall single measure of success. The remaining data on the five girth measures can be used in a parallel way to answer the same questions that were asked using the weight measure. We wish to test the null hypothesis that there is no difference between experimental and control groups at the three-month follow-up point, controlling for the potentially distorting effects of 'status' and 'clinic' influences. Clearly we can test each girth measure separately in a univariate analysis of variance identical to the analysis of the percentage overweight measure. However, the answers to the five separate questions on the five girth measures are of little use, since they are intended as a set of measures that jointly assess the single phenomenon in which we are interested. (True, some individuals may be concerned particularly with one aspect of their anatomy, but that is not the concern of the general study.) A multivariate analysis of variance is required to answer the general question about the groups of experimental and control subjects: 'do they differ?'.

In Table C.5 the vectors of means (see Section 3.1) are presented for all five variables. This is the basic multivariate pattern to be analysed. Along with the measures are shown the numbers of participating subjects, as in Table C.2. These are generally fewer than in the univariate analysis, as a result of occasional missing values in the data set: the multivariate analysis requires a complete set of measurements on each individual, and those with incomplete information have been excluded from the analysis (see Section 3.1).

Table C.6 reproduces the analysis of variance of Table C.3, but for the multivariate set of data. For each comparison tested in the univariate analysis of Table C.3, there is a corresponding entry in Table C.6 (some of the information is not shown, for the sake of brevity, since it will be more pertinently displayed in Table C.7). At the top of the table, instead of the within-cells sums of squares in Table C.3, we now have the within-cells sums of squares for each of the five variables (the diagonal entries), and the sums of cross-products between the variables. The resulting figures form the sums of products matrix for within-cells

Table C.5 Means and numbers in the cells of the clinic by status by condition cross-classification: the five girth percentage change measures at the three-month follow-up point.

	No.	Bust	Waist	Hips	Thigh	Arm
Status: *Experienced*						
Experimental:						
Clinic A	6	−4.7	−9.2	−4.9	−6.8	−8.8
B	5	−0.9	−1.8	−2.0	−2.2	−2.5
Control:						
Clinic A	10	−4.8	−5.6	−4.2	−8.1	−7.0
B	4	−0.9	−0.3	−1.8	−4.6	−1.0
Status: *Novice*						
Experimental:						
Clinic A	12	−3.0	−5.1	−2.6	−1.7	−6.2
B	7	−1.9	−0.7	−0.7	−2.6	−0.7
Control:						
Clinic A	6	−2.3	−5.8	−1.8	−3.3	−8.2
B	13	−1.1	−0.6	−0.8	−1.5	−4.2

Slimming clinics

Table C.6 Analysis of variance for the data from Table C.1.

Source	d.f.	Sums of products matrices					Roy's stat.
		Bust	Waist	Hips	Thigh	Arm	
Within-cell	55	491	—	—	—	—	
Error		418	1 286	—	—	—	
matrix **E**		461	811	1 042	—	—	—
		148	173	327	2 049	—	
		324	664	711	84	1 612	
Two-factor	3	19.8	—	—	—	—	
interactions		28.9	66.6	—	—	—	
Hypothesis		8.1	9.6	6.9	—	—	.150
matrix **H**		−3.3	−11.4	12.6	54.5	—	(sig .750)
		37.6	84.9	9.9	−25.1	110.5	
Three-factor	1	11.2	—	—	—	—	
interactions		−2.6	0.6	—	—	—	
Hypothesis		1.4	−0.3	0.2	—	—	.047
matrix **H**		−11.7	2.7	−1.4	12.2	—	(sig .774)
		7.5	−1.7	0.9	−7.8	5.1	

error, denoted as **E** (see Section 4.3). Similarly, the remaining entries in the table are hypothesis sums of products matrices (denoted **H** in Section 4.3) corresponding to the hypothesis sums of squares for the three-factor and two-factor interactions in Table C.3. The sums of products matrices for the main effects of the factors have not been printed (they are not used in this analysis) simply to avoid too cumbersome a table.

Starting at the foot of Table C.6 as before, the **H** matrix for the three-factor interaction is used to test the null hypothesis that this interaction is zero. Taking it in ratio to the **E** matrix at the top of the table and finding the eigenvalues (Chapter 4) allows the computation of Roy's test statistic, at a value of 0.047 (Section 4.3). This does not reach the 5% significance level and we conclude that, as in the univariate analysis in Table C.3, this interaction is not required to explain the data pattern. Similarly, the entry for the two-factor interactions, when compared with the within-cells error matrices, also fails to achieve significance, with the conclusion that these interactions are not required.

Continuing now down Table C.7 to examine the main effects of the factors, it should first be noted that for the reasons given in the analysis of Table C.4, the variables have been adjusted by a value of 100; the sums of products computations are for the 'unique' (or additional) effect of each factor, controlling for the possible effects of the other two factors. Thus Table C.7 is an exact parallel of Table C.4. The only main effect to achieve significance at the 5% point is the

134

'clinic' effect. Between clinic A and clinic B, after controlling for the influences of disparate numbers of novice and experienced slimmers, and for the disparate numbers in control and experimental groups, there is a tendency for clinic A to yield a greater reduction in its members' girth. The test of the effect of central interest – the comparison of experimental with control groups – is not significant. Finally, we can see at the top of the table that the entry for overall level shows a highly significant result. In other words, the null hypothesis that the overall level of the response variables is zero must be firmly rejected. Again, this overall level is an average of two distinct levels, that of clinic A and that of clinic B. We conclude that in general there has been an overall reduction in girth during the follow-up period; this general reduction has superimposed on it the differential reduction between the two clinics.

Table C.7 Analysis of variance for the revised data from Table C.5 ('unique' sums of cross-products computations).

Source	d.f.	Sums of products matrices					Roy's stat.
		Bust	Waist	Hips	Thigh	Arm	
Within-cell	55	491	—	—	—	—	
Error		418	1 286	—	—	—	
matrix E		461	811	1 042	—	—	—
		148	173	327	2 049	—	
		324	664	711	84	1 612	
Constant	1	292	—	—	—	—	
Hypothesis		474	758	—	—	—	
matrix H		317	511	344	—	—	.660
		494	795	536	834	—	(sig .000)
		674	1 085	731	1 138	1 553	
Clinic	1	161.0	—	—	—	—	
Hypothesis		269.4	451.1	—	—	—	
matrix H		93.0	155.7	53.8	—	—	.475
		97.9	164.0	56.6	59.6	—	(sig .000)
		232.0	421.8	145.6	153.3	394.4	
Status	1	29.6	—	—	—	—	
Hypothesis		26.1	23.0	—	—	—	
matrix H		35.6	31.4	42.8	—	—	.124
		74.5	65.8	89.7	187.9	—	(sig .226)
		1.6	1.4	2.0	4.1	0.1	
Condition	1	10.6	—	—	—	—	
Hypothesis		−6.8	4.3	—	—	—	
matrix H		−3.1	2.0	0.9	—	—	.110
		15.7	−10.0	−4.5	23.1	—	(sig .278)
		18.3	−11.7	−5.3	26.9	31.5	

Slimming clinics

Note that we have not shown necessarily that both clinics have been effective. The above results are consistent with one clinic achieving no improvement and the other clinic achieving a substantial drop in girth: this too would imply an overall level that had dropped. This issue must be resolved by testing each clinic's overall level separately. Table C.8 presents this analysis, which can be regarded as an analysis of two separate experiments – one in clinic A and one in clinic B – except that the error variation within each is pooled into a common sums of products (the **E** matrix). Alternatively, it can be regarded as treating the analysis as a nested one, in which some effects of the between-subjects design, in this case, only the constant or mean, are nested within the levels of the 'clinic'

Table C.8 Tests for the means of each clinic (see Table C.5).

Source	d.f.	Sums of products matrices					Roy's stat.
		Bust	Waist	Hips	Thigh	Arm	
Within-cell	55	491	—	—	—	—	
Error		418	1 286	—	—	—	
matrix **E**		461	811	1 042	—	—	—
		148	173	327	2 049	—	
		324	664	711	84	1 612	
Constant	1	435	—	—	—	—	
within 'A'		692	1 101	—	—	—	
matrix **H**		383	610	338	—	—	.724
		808	969	536	852	—	(sig .000)
		825	1 314	728	1 156	1 568	
Constant	1	3	—	—	—	—	
within 'B'		3	4	—	—	—	
matrix **H**		11	12	38	—	—	.156
		21	25	78	157	—	(sig .128)
		21	24	76	154	151	

factor. The corresponding tests for the mean of each clinic are made by comparing the respective hypothesis matrices with the same error matrix. We can see that, whereas clinic A has achieved a significant improvement, clinic B has made very little change. (The discrepancy is summarized in the value of the canonical variable in each of the two clinics, which we return to below in Table C.9.) The researchers found this result from the study difficult to explain, particularly since the 'overweight' analysis gave different results, save for the possibility that the regimes in the two clinics were in substance rather different and not equally efficacious.

Before proceeding, let us consider the reasons for using tests based on Roy's statistic and for using a 5% level of significance in the preceding analyses. The several test statistics available (see Section 4.3) all give an equivalent test when the hypothesis matrix has only one degree of freedom. Since all factors in this study have only two categories, all the tested effects satisfy this requirement, apart from the test of the pooled two-factor interactions. This test, with three degrees of freedom for the hypothesis, is therefore the only one in which it matters which statistic we choose. Given the nature of the measurements, the sort of departures from the null hypothesis it would be reasonable to expect, if any, are shifts in all variables simultaneously in the same direction. This leads us to expect a basically unidimensional distribution of the group means. Such a concentrated noncentrality structure (see Section 4.5) is in general most powerfully detected by Roy's statistic.

In choosing a significance level we are, of course, at liberty to choose any level we like and need not have made all the tests at the same level, provided we are aware of the implications for type I and type II errors (see Chapter 4). We know that when several variables are being tested simultaneously it is sometimes advisable to use a less stringent overall significance level, in order to bring the individual type I error for each variable more in line with the sort of level we would use in a univariate test. Instead of testing the main effects at the 5% significance level, if for instance we use the rule of thumb given in Section 5.2, we would have tested the overall hypothesis on all five responses at the $1 - (1 - .05)^5 = 23\%$ level. Note that the effect of 'status' would then have been just significant (in fact two of the five measures would be more or less significant in individual tests). However, here we are concerned with a question relating to all five variables simultaneously, so an overall 5% level was chosen for testing this overall null hypothesis – the one in which we are interested – regardless of individual test levels of the component hypotheses.

Two comments are in order here about the choice of variables in this multivariate analysis. It would have been possible to include the overweight variable used in the univariate analysis with the girth variables and perform the analysis of variance on all six variables simultaneously. The reason for not doing so was that the researchers preferred to keep distinct the two concepts of weight and shape, rather than combine them into a single concept of 'success'. The relationship between the two concepts is further explored in a subsequent analysis.

The second comment concerns the scale of measurement of the five girth variables used in the analysis. Each of the set is measured on a single scale, percentage improvement. This commensuration – a score on one variable is directly comparable with a score on another variable – is not necessary in the analysis performed here. It is occasioned simply by the desirability of using this type of scale for physical measurements. The multivariate test statistic that was used, or indeed those which could have been used, are all scale-free and yield the same answer if, for instance, any of the variables had been multiplied by 1000 or measured as a proportion instead of as a percentage (see Section 4.3). It is only when the researcher uses a prespecified transformation – contrasts and linear combinations – of the variables that care must be taken over their relative scales of measurements (see Section 3.8).

In the present analysis interest attaches to measuring the overall reduction in girth. A single summary score of the general reduction would be useful in characterizing the individuals who participated. The test of the constant term in Table C.7 shows that there *is* a significant change in the general levels of girth for the five variables and (in conjunction with Table C.8) that this occurred for those attending clinic A. The canonical variate associated with the test of the constant term of the between-subjects design (in each clinic) gives the summary score we desire (see Section 3.10). Since there is only one degree of freedom in each test, there can only be one canonical variate, and its coefficients are given in Table C.9. The significance test for the constant term in clinic A has associated with it a canonical variate with a pattern of coefficients which sets the 'hips' measurement against all the others. An informal interpretation of this summary variable therefore suggests that this particular bodily dimension is more difficult to change than the others.

Table C.9 Coefficients for the canonical variables associated with the tests of the constant term for each clinic in Table C.8.

	Bust	Waist	Hips	Thigh	Arm
Clinic A	−.587	−.652	1.067	−.614	−.652
Clinic B	.082	.257	.035	−.681	−.842

C.3

So far the two sets of measurements, overweight and girth, have been analysed separately since, as was stated at the outset, the researchers had two different concepts in mind. In the following analysis the two are linked to look at their relationship with each other. It is clearly possible to regard overweight as acting as a summary of the other variables that were measured. Just how effective a summary overall weight can be considered to be, was a subsidiary question in the research. An alternative way of formulating this question is to ask whether we need the girth measurements to explain the various differences in the response variables, or whether that information is contained satisfactorily in the single overweight measurement. A partial answer to this question is found in the analyses already performed, since different results were obtained in the univariate analysis of overweight, as compared with those in the multivariate analysis of girths. In the former analysis only 'status' yielded a significant response difference, whereas in the latter analysis only 'clinic' seemed to yield an effect. However, the formal answer to this question is provided by a multivariate analysis of covariance, in which the analysis on girth measurements is repeated, but this time the overweight measurement is used as a covariate. This is equivalent to asking whether the girth measurements add anything to the discrimination between the groups – that is, do the measures show any differences over and above differences in them that would be predicted from their relationship with the overweight measure?

Table C.10 presents this analysis of covariance. Note first that the numbers of individuals comprising this analysis are fewer than in the simple multivariate analysis of variance. This is a result of occasional missing values in these data; for this analysis we need to have a complete set of data for both the girth measurements and also the overweight measurement. Those individuals not meeting this extra requirement are necessarily excluded from the analysis. The analysis presented in Table C.10 is an extension of that presented in Table C.7, which used only the main effects of the factors as a model. It was thought justifiable to discard the three-factor and two-factor interactions a priori, because of the very small effects associated with them in the previous analysis. The covariance analysis controls for the effect of the covariate, overweight, by means of a regression analysis. We see in Table C.10, therefore, not only the hypothesis matrices for the main effects, but also one for the effect of regression.

The first entry in Table C.10 is the within-cells error matrix, now adjusted for the regression of the response variables on the covariate. It is therefore not simply the within-cells error matrix of Table C.7, but rather the matrix of errors about the regression within each cell. We can see that the entries for the sums of squares in this adjusted error matrix are smaller than the corresponding entries in Table C.7. In direct analogy with simple regression, these sums of squares and products are the residual sums of squares about the regressions of the responses on the covariate, overweight. By contrast, the entries in Table C.7 correspond to the familiar total sums of squares in a regression analysis (that is, the sums of squares about the regression plus the sums of squares due to the regression). The degrees of freedom associated with this matrix are one fewer because of the regression and a further two smaller because of the reduction in the number of observations mentioned above.

The second entry in Table C.10 is the hypothesis matrix for the regression of the girth measures on the covariate – the part 'removed' from the preceding error matrix. This is tested in the same way as any other hypothesis matrix, that is, by taking it in ratio to the error matrix: it parallels exactly the ordinary test for a regression coefficient in simple univariate regression. The hypothesis matrix has a structure identical to the error matrix: it presents not only the regression sums of squares for each of the five girth measurements regressed individually on the predictor variable 'overweight', but also the cross-product terms between these regressions. In assessing the regression effect on the response variables we

Slimming clinics

Table C.10 Analysis of covariance for the revised data from Table C.5, with the data from Table C.2 used as a covariate (unique sums of squares calculations).

Source	d.f.	Sums of products matrices					Roy's statistic
		Bust	Waist	Hips	Thigh	Arm	
Within-cells	52	300	—	—	—	—	
Adj. error		200	1 002	—	—	—	
matrix **E**		178	475	617	—	—	—
		−43	−54	231	1 653	—	
		152	442	445	−77	1 434	
Regression	1	189	—	—	—	—	
Hypothesis		212	239	—	—	—	.499
matrix **H**		279	314	412	—	—	(sig .000)
		203	228	300	218	—	
		166	187	245	178	146	
Constant	1	42	—	—	—	—	
Hypothesis		84	171	—	—	—	.502
matrix **H**		31	63	23	—	—	(sig .000)
		123	249	91	362	—	
		151	306	112	445	548	
Clinic	1	104.0	—	—	—	—	
Hypothesis		175.4	295.8	—	—	—	.465
matrix **H**		36.7	61.9	13.0	—	—	(sig .000)
		65.7	110.9	23.2	41.6	—	
		167.7	282.8	59.2	106.0	270.4	
Status	1	0.6	—	—	—	—	
Hypothesis		0.4	0.3	—	—	—	.069
matrix **H**		0.3	0.2	0.1	—	—	(sig .615)
		7.3	5.0	3.4	96.7	—	
		−2.2	−1.5	−1.1	−19.7	9.1	
Condition	1	0.6	—	—	—	—	
Hypothesis		−4.3	32.4	—	—	—	.307
matrix **H**		−3.9	29.0	26.0	—	—	(sig .345)
		0.4	−3.3	−2.9	0.3	—	
		2.0	−15.2	−13.7	1.5	7.2	

could test each of the five regression coefficients separately. However, in this study we are interested only in the joint response of the girth variables, so a multivariate simultaneous test of the regression coefficients is carried out, using the entire hypothesis matrix (including the off-diagonal cross-product terms) shown in the table. The usual test statistics are suitable for this purpose and Roy's largest root attains a value of .499, which is significant beyond the 0.001

level. Thus, a highly significant regression relationship exists between girth and overweight, that is, girth can be strongly predicted from overweight. Note that these regressions have been carried out within each cell of the design and the results pooled into a single matrix. Given the assumption that the regression slopes are the same for each cell, this test has shown that on average the relationship holds within each cell of the design.

Before continuing our examination of Table C.10 we should note that Table C.11 presents the five separate regression coefficients. The individual significance levels of these regression coefficients, it was argued, are not relevant and have not been shown. The canonical correlation between the five variables

Table C.11 Regression coefficients from the analysis at covariance in Table C.10 (the five girth response variables regressed on the covariate, overweight).

Bust	Waist	Hips	Thigh	Arm
.34	.38	.50	.36	.30

Eigenvalue .9955		Canonical Corrn. .706

and the predictor, overweight, is .71, indicating the strength of the predictive relationship for girth measurements in general that can be achieved from the single overweight variable. The canonical relation here represents the multiple correlation between the predictor variable overweight and that combination of the five response variables which is most easily predicted. In the usual manner, we may take this multiple correlation squared – a value of .50 – as the proportion of variance explained by the regression.

Returning to Table C.10, the entries corresponding to the main effect of the 'clinic', 'status' and 'condition' factors have the same interpretation as in Table C.7, except that we must bear in mind that these comparisons are being assessed now after controlling for the effect of the covariate. The question we are asking is whether they show differences above and beyond those that would be predicted from differences in overweight (see Section 3.8). In testing to see whether these comparisons are making significant contributions to the response pattern it is necessary to control each main effect for the influences of the other two main effects, partly because of the discrepant patterns of response associated with the factors and the variables. Table C.10 therefore presents computations of the 'unique' sums of products attributable to the factors, just as did Table C.7. Although the sums of products in these hypotheses matrices are somewhat reduced when compared with Table C.7, there is little change in the resulting substantive interpretation. The comparison of central interest, the factor 'condition', gives a non-significant value of Roy's largest root statistic, as does

Slimming clinics

'status'. The effect of the 'clinic' factor is still highly significant, as is the test on the constant term.

Table C.12 gives the size of the estimated effect of 'clinic' and the value of the constant term for the five girth measures and for the canonical variable associated with the test. The lower part of Table C.12 gives the same figures for the analysis of variance without using a covariate, that is, for the analysis of Table C.7. We can see that there is little change in the size of the effects on the canonical variables.

Table C.12 The estimated value of the constant term and the effects of the 'clinic' factor for the data in Table C.2.

	Bust	Waist	Hips	Thigh	Arm	Canonical vbl.
Controlling for the effect of the covariate (see Table C.10):						
Constant	−1.0	−2.0	−0.7	−3.0	−3.6	−1.13
Clinic	−1.4	−2.3	−0.5	−0.9	−2.2	−0.91
Without considering the covariate (see Table C.7):						
Constant	−2.2	−3.4	−2.4	−3.9	−4.7	−1.27
Clinic	−1.6	−2.7	−0.9	−0.9	−2.6	−0.91

In conclusion, it can be said that although there is a strong regression relationship between the girth measures and the overweight measure, there is still additional information in the girth measures for assessing individual performance, with respect to both overall improvement and clinic differentials. We cannot safely presume that overweight is an adequate summary of the various aspects of performance in this study.

C.4

The final aspect of this study which is reported here is the investigation of how people fared over a longer period of time. There was considerable interest in checking on the long-term maintenance of any short-term change of eating habits that was achieved. The long-term behaviour of the group was measured only on the percentage overweight variable and this was taken at three time points, 9 weeks, 3 months and 1 year. Difficulties in tracing individuals over a period always occur in longitudinal group studies, and in this particular study only 64 of the original 89 individuals were successfully traced and weighed. In fact, most of the drop-out occurred between the finish of the course at 9 weeks and the 3-month follow-up point. The vast majority of those not measured at one year consisted of people who were untraceable, with only a few refusals.

Omitting such groups of people from the analysis can lead to bias in the results, in particular if there are many refusals, since people might have stronger reasons for refusing if they feel that they have failed in the course. To proceed it is necessary to assume that these omitted individuals are a random selection of individuals initially recruited to the trial. Checks can be made to see if they differ on any measurements taken prior to their dropping out of the study, but these must be considered as minimal checks which cannot guarantee the randomness of the self-selected group of individuals who were not traced. The results of this analysis must therefore be considered to be of dubious validity, if the assumption of randomness of drop-out is violated. Table C.13 shows in detail how many subjects were available for each analysis. Note again that this is fewer than the number of subjects who were actually measured at, say, one year. To be available for the analysis of data over the entire year of follow-up, an individual must be measured at all three time points.

Some further cause for concern in this particular analysis comes from the distribution of two of the response variables. It was thought that the numbers of observations within each cell were, in general, too small to indulge in any serious testing of normality. However, by plotting the deviations of each observation from its cell mean and pooling these plots into one diagram, two or three outlying observations were revealed. If these observations were suspected of being generated by some mechanisms which would enable us to define them as a different, deviant population, it would have been possible to exclude them from the analysis. In this study there were no grounds for such a line of reasoning, so

Table C.13 Numbers of subjects available for analysis at the 9-week, 3-month and 1-year follow-ups.

Condition	Status: Novice		Status: Experienced	
	Clinic A	Clinic B	Clinic A	Clinic B
Measured at 9 weeks				
Experimental	9	9	12	9
Control	10	6	4	14
3-month follow-up analysis				
Experimental	5	5	12	9
Control	11	4	6	13
1-year follow-up analysis				
Experimental	4	3	4	6
Control	8	3	3	9

Slimming clinics

Table C.14 Means and numbers in the cells of the clinic by status by condition cross-classification: change in percentage overweight (from baseline measure) at the three follow-up points.

	No.	9 weeks	3 months	1 year
Status: Experienced				
Experimental:				
Clinic A	4	−5.0	−5.5	−4.4
B	5	−5.9	−2.5	−2.3
Control:				
Clinic A	8	−8.1	−7.5	−6.9
B	3	−6.3	−4.7	−5.2
Status: Novice				
Experimental:				
Clinic A	10	−3.3	−0.7	−0.5
B	6	−2.3	−0.0	−1.4
Control:				
Clinic A	3	−1.0	−0.6	−1.0
B	9	−6.0	−3.0	−1.1

instead these observations were included in the analysis, and appeal made to the robustness properties of least squares estimation and test procedures. Nevertheless, the approximate nature of the ensuing tests should be borne in mind.

The between-subjects design in this analysis is, as before, a three-factor crossed design, but now there are only three response variables: percentage weight loss at 9 weeks, 3 months and 1 year. Table C.14 shows the vector of mean values for each of the cells in the between-subjects design and the numbers of individuals in these cells. Again a decision was made to look at only the main effects of the factors. A more rigorous analysis would test for the existence of interaction effects rather than assume for simplicity of exposition, as is done here, that these interactions are zero.

There are a number of questions which could be asked of these data. In this long-term follow-up one of the researchers' central questions was to ask whether any weight loss was shown on average throughout the follow-up year. Secondly, it was of interest to ask if there were any *changes* throughout the year and, if so, when these changes occurred. Logically, these questions should be asked in the reverse order, since in many circumstances dramatic changes in weight throughout the year would make the average level of weight loss irrelevant. We have remarked before that when response variables are commensurate it might be possible to pre-specify particular linear combinations of these variables which will be of especial relevance (see Section 3.4). In this study, clearly the difference between the 9-week and 3-month follow-up measurements is highly relevant, as too is the difference between the 3-month and 1-year measurements. These are

144

precisely the changes through the year, mentioned above, in which the researchers were particularly interested. These contrasts can be written in the matrix form:

$$\begin{matrix} 1 & -1 & 0 \\ 0 & 1 & -1 \end{matrix}$$

In addition to these two contrasts, the overall mean of the variables is also of interest, in that it can be used to answer the question about the general over-weight level throughout the year. Thus we need to add to the two preceding contrasts the linear combination:

$$\begin{matrix} 1 & 1 & 1 \end{matrix}$$

Note that these linear combinations are exactly those that we would use in analysing any factor such as those we have already seen in the between-subjects design. Here we need to treat these three overweight measures not simply as a set of three variables to be analysed simultaneously, but as a factor 'time' at three levels. This factor, a between-variables effect of the factor 'time', compares differences in weight loss at different follow-up points; the overall level of observation provides us with the average performance throughout the year (see Section 3.5). Note that from the two contrasts stipulated in the preceding matrix, we can derive the difference between levels 1 and 3 adding the first row to the second. Thus these contrasts provide a complete comparison (see Section 2.3) for the factor – indeed, these are the successive difference contrasts given in Section 2.15, para. 3. In this analysis they are being used as a transformation matrix applied to the variables, providing us with a useful set of contrasts for a between-variables factor.

Our analysis, then, is of the by now familiar between-subjects three-factor design, the two-by-two-by-two cross-classification. The between-variables design is one factor, 'time', in which the comparison for the main effect is represented by the pair of contrasts giving successive differences between follow-up measures. Note that the contrasts used in the between-subjects design, as in the preceding analyses, can only be the difference between the two levels since each between-subjects factor has only one degree of freedom.

Table C.15 presents the analysis for this design. The transformation matrix used with the between-variables factor produces two new variables to describe the main effect, which has two degrees of freedom (two change scores), and one new variable indicating the overall level throughout the year of follow-up, having one degree of freedom. The analysis of the main effect and the analysis of the overall level can be carried out separately and are presented in Table C.15(i) and (ii), where the between-subjects design is analysed with respect to each of these two sets of variables. The analysis of the main effects of 'time' is a multivariate analysis with two response variables, labelled 'Change 1' and 'Change 2'. The analysis of the overall level, having only one variable in it

Slimming clinics

Table C.15 Analysis of variance for repeated measures data in Table C.14 (unique sums of squares computations in the between-subjects design).

(i) Constant term (overall mean level) of between-variables design, 'Follow-up'.

Source	d.f.	Sums of squares Follow-up	F-statistic
Within cells	38	1 127.5	—
Constant	1	600.8	20.25 (sig .001)
Clinic	1	3.6	0.12 (sig .730)
Status	1	136.8	4.61 (sig .038)
Condition	1	45.8	1.54 (sig .222)

(ii) Main effect (successive difference contrasts) of between-variables factor time, 'Change 1' and 'Change 2'.

Source	d.f.	Sums of products matrices Change 1	Change 2	Roy's statistic
Within-cells	38	493.4	—	—
Error matrix **E**		−92.4	1 062.0	
Constant	1	148.1	—	.256
Matrix **H**		55.0	20.4	(sig .004)
Clinic	1	11.5	—	.035
Matrix **H**		10.4	9.4	(sig .517)
Status	1	17.7	—	.036
Matrix **H**		1.8	0.2	(sig .508)
Condition	1	0.4	—	.001
Matrix **H**		0.2	0.1	(not sig.)

(labelled 'Follow-up'), reduces to the familiar univariate analysis of variance. Table C.15 does not present the analyses for the between-groups interaction terms, these having been excluded a priori, as stated above – a more rigorous analysis would include these interaction sums of products matrices with the three main effect sums of products matrices for clinic, status and condition. The computations in Table C.15 are of the 'unique' or additional sums of products for each between-subjects factor, as in the analyses in all of the preceding sections we wish to control for the possible effects of other factors when assessing each individual comparison.

We look first at the multivariate analysis in Table C.15(ii). By comparing each of the hypothesis matrices in turn with the within-cells error matrix – that is, by taking the ratio $\mathbf{H} \cdot \mathbf{E}^{-1}$ – we can compute Roy's largest root statistic for the multivariate test of hypothesis. The entry at the foot of the table is the hypothesis matrix for the main effect of 'condition'. The null hypothesis being tested is that each of the two new variables – that is, Change 1 and Change 2 in the 'time' factor – has the same mean value in the 'novice' level of condition as it does in the 'experienced' level of condition. More precisely, it is an overall or simultaneous hypothesis that the response mean vector is the same for all levels of the factor. It does not matter which of the multivariate statistics given in Section 4.3 we use to test this hypothesis, since all give equivalent results when, as here, the hypothesis has only one degree of freedom. However, we would in general benefit from considering the sort of departures of null hypothesis that might be expected: we would probably expect a diffuse noncentrality structure in this example (see Section 4.5) and if there were more than one degree of freedom in the hypothesis, the Pillai–Bartlett statistic might be more powerful. Against a 5% significance test level, the effect of 'condition' is not significant and we conclude that the hypothesis of equal changes in weight for the two conditions is not rejected by evidence from these data. The phenomenon of equal changes in different groups throughout time is sometimes called by psychologists a 'parallel profile'. The remaining two entries in the table show, similarly, that the main effects of 'status' and 'clinic' are also non-significant at the 5% level.

Although this analysis shows that none of the three factors' effects is significant when controlling for the other two, it is conceivable that, in unbalanced designs, one or more of them do have a simple effect when the other factors are not controlled for. If, for instance, most people are either 'experienced' control subjects or 'novice' experimental subjects (and few belong to the other two combinations of levels) it may be possible to discern an effect on the response variable, but impossible to allocate its cause definitely to one factor and not the other. It is therefore necessary to check on these possibilties by further testing of the factors, controlling for only one or for neither of the other two. These analyses are not presented here, but no significant effects were revealed by these procedures.

Continuing with the tests of the two change scores in Table C.15(ii), we see that the hypothesis matrix for the constant, or grand mean, of the between-subjects design is significant at the 5% level, so the compound null hypothesis that the two change contrasts have zero means across all subjects is rejected. Thus although the time profiles for the groups are parallel, these profiles are not null, and changes between one follow-up point and the next have occurred. The question of identifying which of the changes is significant (or whether both are) will be held over until a later analysis.

We turn now, in Table C.15(i), to the analysis of the overall level of the three follow-up measures, that is, the analysis of the variable 'Follow-up'. This is the familiar univariate analysis of variance table, since we are now dealing with only one variable. Note that the significance tests made in this upper part of the table are the usual F-tests formed by comparing the hypothesis mean square with the

147

Slimming clinics

within-cells error mean square. Again using a 5% significance level we see that of the three between-subject factors only one, 'status', shows a significant effect. Thus the null hypothesis that the 'novice' and 'experienced' groups have the same overall levels of weight loss is rejected. Finally, turning to the constant for the between-subjects design – that is, the grand mean – we note in passing that it is significantly different from zero. Thus the null hypothesis that the overall weight loss for the entire sample throughout the year is zero is rejected. Again, however, this grand mean is not of particular interest since we have shown that two subsets of people in the sample – 'novice' and 'experienced' – differ in their responses, making the mean for the whole sample a difficult figure to interpret. These two effects (the 'constant' and 'status') constitute an adequate model needed to account for the response 'Follow-up'.

Table C.16 Means and numbers in the cells of the status classification: change in percentage overweight (from baseline measure) at the three follow-up points.

	No.	9 weeks	3 months	1 year	Year-long
Status: Experienced	20	−6.7	−5.7	−5.3	−5.9
Status: Novice	28	−3.7	−1.1	−0.3	−1.7
Grand mean	48	−4.9	−2.9	−2.3	−3.4

These results are presented in Table C.16. It shows, for the 'Follow-up' variable, the estimated values of the 'status' comparison and of the grand mean (that is, of the year-long average weight loss for the whole sample). It also shows the mean weight loss at each follow-up point for the two 'status' groups and for the sample as a whole. In so doing it is presenting a condensation or summary of the original data mean values given in Table C.1.

Now we take up the question held over from an earlier paragraph, 'what changes have taken place through the year?'. In Table C.15(ii) the hypothesis matrix for the compound test of the mean vector for Change 1 and Change 2 showed that changes existed. The hypothesis matrix associated with the constant term in the between-variables design was used in that table to test, as in Table C.7, the null hypothesis that the mean vector is equal to zero; it yielded a significant result, indicating that not all the means are zero. Having rejected the overall hypothesis in this way, we may now turn to its constituent hypotheses to see which of these is to be rejected, that is, on which of the two variables does the change occur (or do both change, or does any combination of them change)? When hunting through these constituent hypotheses to see where the overall hypothesis fails (see Section 3.2) we should control the overall error rate by using a simultaneous test procedure (see Sections 5.1 and 5.2).

In the present analysis we may use an extremely simplified form of Gabriel's procedure, in which there are no groups but only variables to identify (see Section 5.5). We know from the property of consonance of this procedure with Roy's statistic that one such combination of the two responses is significant. The most significant is the first (and only, in this one degree of freedom hypothesis) canonical variable associated with the test: it is the most effective summary of the ratio of hypothesis to error matrix (see Section 3.10). In this analysis the researchers were particularly interested in the two very simple 'combinations' of the responses, Change 1 and Change 2 themselves. There was also particular interest in the combination formed by the sum of these two variables, since it gives, obviously, the change from the first time point (9 weeks) to the last (1 year). The test statistics for each of the three hypotheses – that the changes referred to are zero – are shown in Table C.17. Comparing these with the critical level appropriate to the *overall* test statistic at the 5% level, as Gabriel's procedure requires, shows that only the first is significant at this level. Note that all possible combinations of these two variables are covered by the same probability statement.

In conclusion, we can say that an average weight reduction throughout the year has been achieved, and that this is different for the 'novice' and 'experienced' groups. Furthermore, there have been changes in weight loss about this average from one follow-up to the next, but there is no evidence that these changes differ from group to group.

Table C.17 Tests on combinations of the response variables for the significant between-groups 'constant' effect in Table C.15.

	Change 1	*Change 2*	*Change 1 + Change 2*
Error Sum of squares	493.4	1,062.0	1,370.6
Constant Hypothesis Sum of squares	148.1	20.4	278.5
Roy's statistic (*F*-statistic)	.300	.019	.203

Table C.18 Percentage improvement scores for the (revised) three follow-up overweight measures and five girth measures used in the example.

Cond		Stat	Clin	OW9	OW3	OW1	B	W	H	T	A
1	1	2	1	−2.42	.81	.81	−3.00	−4.82	−5.71	−1.69	.00
2	1	2	1	.00	2.38	.79	−.95	−2.33	−.94	−5.08	.00
3	1	2	1	−2.74	2.74	.00	−1.82	−.98	.00	5.45	−2.86
4	1	2	1	−.84	3.36	−.84	−3.13	−5.13	.00	1.89	−6.25
5	1	2	1	.00	2.10	−2.10	−4.63	−3.06	−2.52	4.69	−15.38
6	1	2	1	−9.17	−.83	.	−2.15	−7.32	−9.01	1.64	−11.43
7	1	1	1	.00	−14.67	.	−.98	−9.09	−1.85	−9.84	−5.88
8	1	1	1	−1.85	−1.85	−4.63	−2.11	−5.19	−1.92	−8.47	−3.23
9	1	2	1	−3.05	−3.05	−3.05	−1.92	−2.38	−4.39	−9.68	−6.45
10	1	2	1	−9.40	−5.98	−6.84	−6.82	−12.16	−5.94	−8.47	−15.63
11	1	2	1	.	−3.64	.	−1.23	−7.35	−2.13	−5.45	−7.14
12	1	2	1	−7.38	−6.56	−.82	−3.00	.00	−1.98	.00	−6.06
13	1	2	1	−3.60	−1.44	1.44	−5.71	−8.51	−1.72	−1.72	.00
14	1	1	1	−5.98	−8.55	−1.71	−4.35	−8.86	−5.00	3.57	−9.09
15	1	2	1	−3.77	−.94	5.66	−2.17	−7.69	3.06	−1.67	−3.57
16	1	1	1	−12.12	−23.03	−23.03	−13.51	−17.02	−15.70	−18.46	−19.44
17	1	1	1	.00	11.61	11.61	−1.96	−6.59	.00	−13.11	−6.06
18	2	1	1	.00	−3.39	−6.78	.	.	.	.	.
19	2	.	1	.00	−7.14	.89	.	.	.	.	.
20	2	2	1	.00	5.84	−2.19	.	.	.	.	.
21	2	2	1	.00	1.71	.	1.06	.00	4.26	−5.26	.00
22	2	2	1	−1.64	−4.10	−.82	−2.97	−1.25	−2.00	−3.45	−3.23
23	2	1	1	−6.40	−4.00	.	−4.76	−11.83	−6.60	−1.96	−14.71
24	2	1	1	−6.15	−2.31	10.77	−1.98	2.50	2.73	9.52	−8.33
25	2	.	1	−5.74	−4.10	−2.46	−1.11	−2.63	−1.00	−1.56	−9.38
26	2	2	1	−4.44	−5.19	.	−3.85	−2.30	−3.70	−9.38	−14.29
27	2	1	1	−8.11	−3.60	.00	−5.38	−9.09	−7.14	−12.96	−10.34
28	2	1	1	−9.40	−7.69	−7.69	−8.74	−7.06	−7.14	−2.99	−6.06
29	2	1	1	−14.56	−13.92	−21.52	−7.83	−10.68	−7.44	−14.29	−5.56
30	2	2	1	−1.50	.00	.00	.80	−9.52	.00	.00	−11.36
31	2	2	1	−3.74	−2.80	.	−5.68	−10.26	−4.26	.00	−6.90
32	2	1	1	−6.25	−7.64	−13.89	−3.85	5.00	−3.42	−20.90	−6.06
33	2	1	1	−5.52	−7.59	.	−5.45	−8.79	−.94	−1.56	−8.11
34	2	1	1	−3.78	−1.62	.	2.59	3.96	.78	−18.57	.00
35	2	1	1	−13.74	−12.21	−12.21	−4.67	−11.11	−5.61	−7.14	−14.29
36	2	1	1	−6.19	−8.85	−3.54	−7.77	−8.79	−7.69	−9.80	3.70
37	1	1	2	−1.59	2.38	4.76	1.90	6.32	5.83	.00	7.41
38	1	2	2	−.89	−.89	.	.00	.00	−2.88	−3.51	−5.71
39	1	2	2	−.88	−.88	.	2.08	−1.28	−2.06	−3.45	3.33
40	1	2	2	.00	9.09	2.73	5.49	2.53	4.08	.00	.00
41	1	2	2	−1.61	1.61	4.84	4.30	4.94	1.92	.00	.00
42	1	2	2	−1.69	.00	−1.69	.00	.00	.98	−1.59	3.23
43	1	2	2	−5.69	−7.32	.00	−1.08	−1.22	−6.36	−12.70	−3.13
44	1	2	2	.88	.88	6.19	.	.	.	.	.
45	1	2	2	−5.41	−4.50	−3.60	.	.	.	.	.
46	1	1	2	−8.33	−7.58	−6.82	−2.20	−5.13	−7.41	−8.06	−6.67
47	1	2	2	.	−.96	38.46	2.20	.00	−1.02	3.28	6.90
48	1	1	2	−.59	−.59	.	.00	−3.81	−2.42	.00	−7.50
49	1	1	2	−5.07	−.72	.	−6.14	−7.84	−6.84	−4.76	−5.71
50	1	1	2	−7.87	−2.36	−4.72	2.08	1.32	.95	1.67	.00

Neuropeptides and —— Study D
psychiatric illness

D.1

This study arose during an investigation of a proposed link between certain kinds of major psychiatric disease and abnormalities of neuropeptide concentration in the blood. We are grateful to Dr Jeremy Coid (Maudsley Hospital), Dr Bruno Allolio (St Bartholomew's Hospital), and Professor L. Rees (St Bartholomew's Hospital) for permission to use these data.

The subjects being studied fell into two unmatched groups. Five of the subjects were receiving hospital care as either in-patients or day patients, since, due to the severity of their illness (involving self-mutilation) they were unable to function in the community. The other five were either out-patients or had been discharged after treatment. The symptoms of those in this group had disappeared and no self-mutilation had occurred for two months.

The initial primary question of interest was the general one of whether the two groups had a different balance of blood plasma neuropeptide concentrations. If the groups were found to differ then interest would turn to which of the particular four neuropeptides being studied (AC-TH, C-LPH, N-LPH, and Met-enkephalin) or combinations of them were responsible for the difference. Since it was thought that these combinations would depend strongly on age and sex, the experimenter had matched the ten patients on age and sex to ten healthy subjects. (An alternative approach to consider would be to introduce sex as an additional cross-classifying factor and age as a covariate. Unfortunately, data on sex and age were not available at the time of analysis of these data to enable this alternative to be compared with the present analysis.)

The data are shown in Table D.1. A striking feature of these data are the scores of the second control subject, especially on AC-TH and N-LPH. Clearly this subject represents an outlier. In such circumstances, if it is possible, the values should be checked. In our case this was not possible by the stage when the analysis was being carried out and so this subject was dropped from the analysis. The impact of retaining this subject is described below.) Another way to handle extreme values is to transform the response variable by some appropriate formula in the hope that removing skewness from the distribution will remove apparent outliers. Other expert knowledge about the distributions of such chemicals could aid the analyst here. In this example, however, we felt that the

151

Table D.1 Raw data for the neuropeptide study. Four measures on ten patients and ten matched control subjects.

AC-TH		C-LPH		N-LPH		Met-enk	
Pat.	Con.	Pat.	Con.	Pat.	Con.	Pat.	Con.
In-patient group							
56	25	595	225	613	215	140	13
60	104	461	180	463	1074	186	76
17	32	199	510	238	439	106	87
61	24	601	235	559	256	184	38
36	45	373	344	382	381	77	31
Out-patient group							
26	41	189	382	199	384	36	39
39	26	647	364	629	303	35	32
25	24	223	352	304	297	28	77
20	28	193	314	226	287	57	69
28	31	268	296	422	317	12	39

sample size was too small to pursue the merits of such an approach, and that the extremeness of the outlier in any case did not augur well for a successful result.

Since our concern is with differences between the two patient groups, with the controls simply providing baseline measures, we can consider each of the ten patient/control pairs as our basic unit of study. Then we have the four measures taken twice on each unit, but since we are only interested in difference scores we can immediately reduce these eight measures by subtracting the control scores from the matching patient scores. These resulting four difference scores, measured on each matched pair, provide the basic data for the analysis.

As we have discussed in Part 1, to address the general question of whether there are between-group differences it is not appropriate to conduct four separate univariate tests. Instead we should carry out a single multivariate test, a test of the overall null hypothesis that the mean vectors of the two groups are identical. Table D.2 gives the group means for each of the four difference variables, as well as the hypothesis and error sums of products matrices.

Before we use these matrices to derive the value of the test statistic, we should consider at which level we are going to conduct the test. Our sample is not large so we want to avoid imposing too strict a criterion which could fail to detect all but an astronomically large departure from the null hypothesis. For this reason it was decided not to use a 1% level. Conversely, the 10% level seems too lax and we prefer something tighter. In between these levels we have the conventional 5% level, which is what we chose.

In choosing a test statistic note that since we have only two groups, and therefore only one degree of freedom for the hypothesis comparison, all four

Table D.2 Means and sums of products matrices in the neuropeptide study (omitting outlier): the variables are matched-pair differences (see text).

	AC-TH	C-LPH	N-LPH	Met-enk.
Means of variables				
In-patients	11.0	113.5	125.3	84.5
Out-patients	−2.4	−37.6	38.4	−17.6
Overall	3.6	29.6	77.0	27.8
Hypothesis sum of products matrix (2 groups: d.f. = 1)				
	399.0	4 499.4	2 586.2	3 040.3
	.	50 736.0	29 162.3	34 282.9
	.	.	16 762.1	19 705.3
	.	.	.	23 165.4
Error sum of products matrix (d.f. = 7)				
	2 591.2	31 467.8	28 697.8	4 922.8
	.	459 224.2	402 554.7	64 136.2
	.	.	375 787.9	51 142.7
	.	.	.	13 104.2

statistics of Section 4.3 yield the same result. The likelihood ratio method, criterion (2) of Section 4.3, gives a Wilks's lambda of 0.1420 with (1,7) degrees of freedom. The approximate F-value derived from this is 6.045 with (4,4) degrees of freedom. F-tables reveal that this would be significant at the 5.5% level but not at the 5% level. If we interpret this strictly, we are led to the conclusion that we do not have sufficient evidence to reject the null hypothesis of no difference between the mean vectors of the two patient groups. If we do this, concluding that patients and controls have the same means, we may then continue the investigation by seeing whether the mean vector for the whole sample differs from zero, i.e. a test of the constant term in the between-subjects design. This, however, was not a question raised in the initial formulation and may be of no interest.

On the other hand, it is tempting to be less strict in our application of the significance level, and take the observed level as near enough to 5% to be worth pursuing. Since this is a didactic text, however, we shall not do this but instead go on to see how a blind inclusion of the initially excluded outlying point influences the above results, for illustrative purposes.

D.2

If no initial examination of the data had been made we would not have recognized that the second control subject had abnormal values and this subject would have been retained in the analysis. For small data sets such as the current

Neuropeptides and psychiatric illness

Table D.3 Means and sums of products matrices, as in Table D.2, but including the outlier.

	AC-TH	C-LPH	N-LPH	Met-enk.
Means of variables				
In-patients	0.0	147.0	−22.0	89.6
Out-patients	−2.4	−37.6	38.4	−17.6
Overall	−1.2	54.7	8.2	36.0
Hypothesis sum of products matrix (2 groups: d.f. = 1)				
	14.4	1 107.6	−362.4	643.2
	.	85 192.9	−27 874.6	49 472.8
	.	.	9 120.4	−16 187.2
	.	.	.	28 729.6
Error sum of products matrix (d.f. = 8)				
	5 011.2	24 097.8	61 092.8	3 800.8
	.	481 669.2	303 897.2	67 553.2
	.	.	809 439.2	36 123.2
	.	.	.	13 624.4

one, the effect of such an outlying point can be extreme. This is demonstrated in Table D.3, which follows Table D.2 but includes the outlier. Comparing the means in Table D.3 with those in Table D.2, we see that the outlier causes a massive change in mean N-LPH and a smaller but still substantial one in mean AC-TH and C-LPH.

Again using criterion (2) of Section 4.3, we find, from the matrices in Table D.3, that Wilks's lambda is 0.1472 with 1 and 8 degrees of freedom. This yields an $F(4,5)$ approximation of 7.240, easily significant at the 5% level. Thus, if we include the outlier, then even if we adopt a strict 5% level, we end up rejecting the null hypothesis. Outliers usually increase the within-group variance, but here a massive between-groups effect is induced.

Having concluded that the mean vectors of the two groups differ, we can now go on to see if we can identify which of the four variables are chiefly responsible. Clearly a simultaneous test procedure of some kind is appropriate, in order to control for the multiple tests which will be carried out, as discussed in Part 1.

One way to do this might be Gabriel's procedure, as discussed in Section 5.5. Recall that the test statistics are equivalent to one degree of freedom hypotheses. For this we would calculate Wilks's lambda separately for each subset of the variables and compare these values with the 5% critical point of the distribution of the overall multivariate lambda. This latter value is obtained from tables as 0.1940; we should recall that, for this test statistic, the smaller the value, the more significant it is. Note that Gabriel's procedure is general enough to allow comparisons between all subsets of *groups*, as well as of variables. In this analysis there is only one pair of groups to be compared, so this aspect of the procedure is

154

Table D.4 Significance test results for the neuropeptide study: between-group differences. The 5% critical value for Wilks's lambda (d.f. 1,8) is 0.1940.

Hypothesis variables	Wilks's lambda	Sig. level
(i) All sets of three variables		
All variables	0.147	sig.
AC-TH, C-LPH, N-LPH	0.643	not sig.
AC-TH, C-LPH, Met-enk.	0.188	sig.
AC-TH, N-LPH, Met-enk.	0.278	not sig.
C–LPH, N-LPH, Met-enk.	0.193	sig.
(ii) Sets of two variables		
C-LPH, Met-enk.	0.1938	sig.

unnecessary: we look at all combinations of variables in relation to this single between-group difference.

Table D.4(i) lists all subsets of three variables from the original four, resolving the overall (rejected) null hypothesis into constituent null hypotheses. Testing these sub-hypotheses, we see that two of these subsets yield values of Wilks's lambda that are less than the critical value of 0.1940, namely (AC-TH, C-LPH, Met-enkephalin) and (C-LPH, N-LPH, Met-enkephalin). The other two subsets are not significant against this overall critical value. Now the analysis continues by looking at subsets of pairs of measures drawn from these two (rejected) sub-hypotheses. In so doing we can simplify the task by using the property of coherence (see Section 5.3): namely, if any set of variables is not significant as a multivariate set then no subset of them can be significant either. Thus no pair of neuropeptides from the non-significant two sets of three in the table can be significant. The only tests needed are those pairs which occur in the rejected null hypotheses and do not occur in the others – in this case, only (C-LPH, Met-enkephalin). The test for this is given in Table D.4(ii), yielding a lambda 0.1938 which, being less than 0.1940, is significant (at the overall 5% level). Moving finally to tests of single variables, we can immediately see by the same argument of coherence that none of them can be individually significant at the overall level. This follows because they are each contained in at least one of the null hypotheses already tested and not rejected. Note that the property of consonance requires that some linear combination of C-LPH and Met-enkephalin is significant (the canonical variate is) but not that either particular variable is.

The results of this procedure can be summarized in either of the following two ways. The minimal rejected hypothesis list (see Section 5.5) is simply the C-LPH, Met-enkephalin) pair, since its rejection implies the rejection of all

155

those hypotheses marked as significant in Table D.4. The maximal acceptance list contains two entries: (AC-TH, C-LPH, N-LPH) and (AC-TH, N-LPH, Met-enkephalin); any other non-rejected hypothesis is contained in at least one of these two subsets.

We remind the reader that these analyses in the second part of this example are conditional on retaining an apparent outlier in the data. If this is dropped then the initial between-groups test just fails to achieve significance at the 5% level. It might be of interest to see the effect of retaining the outlier and using a log transformation on the original, unmatched data values.

Headache ——————————— Study E

E.1

We are indebted to Clare Philips and Marjan Jahanshahi of the London University Institute of Psychiatry for permission to use these data.

The analysis to be described here was part of a study investigating the effectiveness of different kinds of psychological treatment on the sensitivity of headache sufferers to noise. Each subject was exposed to the following sequence of operations:

- measurement of initial sensitivity scores
- relaxation training
- treatment
- measurement of final sensitivity scores

The sensitivity scores were obtained by listening to a tone which was gradually increasing in volume and having the subjects rate the levels at which the tone became (1) uncomfortable and (2) definitely unpleasant. These levels, denoted U and DU, are the two basic dependent variables in the analysis. Since they are each measured at the beginning and end of the above sequence, we have a total of four variables. Moreover, since interest was primarily in whether treatment groups differed in general sensitivity, and not in the two sensitivity scores U and DU as individual items, it was thought that multivariate analysis was appropriate.

Relaxation training, the second step in the sequence of operations above, was applied to all subjects and comprised two stages:

(a) The subjects were asked to listen to the tone at their definitely unpleasant level for up to two minutes (with the option that they could terminate the exposure if and whenever they chose).

(b) The subjects were then all given instruction on breathing techniques and the use of visual imagery to act as a controlled distraction.

The between-subjects design was a two-by-four factorial design, the first factor being a classification factor of headache type (migraine or tension) and the second a treatment factor with four levels as follows:

T1: subjects in this group listened to the tone again at their initial definitely unpleasant (DU) level for the length of time that they were able to stand it in (a) above.

Table E.1 The raw data for Study E. (Col. (1) = Migraine/Tension, Col. (2) = Treatment; remaining cols. in order U1, DU1, U2, DU2.)

(1)	(2)	(3)	(4)	(5)	(6)	(1)	(2)	(3)	(4)	(5)	(6)
1	3	2.34	5.30	5.80	8.52	2	3	2.57	4.40	3.27	8.64
1	1	2.73	6.85	4.68	6.68	1	3	1.62	3.40	4.03	5.70
2	1	0.37	0.53	0.55	0.84	2	2	4.66	6.82	3.45	6.24
1	3	7.50	9.12	5.70	7.88	1	2	1.12	1.39	1.06	1.78
1	3	4.63	7.21	5.63	6.75	2	4	4.10	7.65	3.36	6.58
1	3	3.60	7.30	4.83	7.32	1	4	2.65	4.88	1.20	3.50
1	2	2.45	3.75	2.50	3.18	1	3	3.58	5.60	6.94	9.16
1	1	2.31	3.25	2.00	3.30	2	3	6.17	15.50	7.54	16.24
1	1	1.38	2.33	2.23	3.98	1	4	0.66	1.00	0.43	0.60
2	3	0.85	1.42	1.37	1.89	2	4	2.08	3.30	2.44	3.47
1	3	1.85	3.25	3.40	4.80	2	2	3.38	8.27	7.07	0.90
1	2	1.90	8.68	2.25	6.70	2	4	3.86	5.94	3.20	4.81
2	3	15.10	16.30	15.50	16.40	1	1	8.94	15.46	3.64	9.00
2	1	0.55	1.10	1.80	3.92	2	1	1.88	4.19	1.79	4.26
2	3	7.42	14.50	8.15	13.30	2	2	2.39	4.60	2.93	5.42
1	4	3.40	5.10	2.80	4.40	2	3	3.62	8.83	5.12	7.71
2	3	1.52	2.35	1.20	2.55	1	4	1.86	4.06	1.78	2.44
2	1	1.85	5.68	7.75	16.10	2	3	2.19	3.94	2.31	3.48
2	1	6.00	9.90	8.25	10.70	2	4	2.08	2.64	1.71	2.99
1	2	1.56	2.92	2.00	2.84	2	4	1.51	2.80	1.24	2.63
2	1	2.95	4.98	3.85	4.75	1	2	1.60	2.83	1.87	3.00
2	1	2.95	3.45	1.75	2.30	1	3	0.75	0.94	0.88	1.45
1	1	2.25	4.40	1.75	4.93	1	2	5.05	10.05	3.02	10.10
1	2	4.22	13.30	12.20	14.10	2	4	3.25	5.09	1.24	2.11
2	2	6.68	9.90	8.52	12.80	1	4	10.36	17.00	11.50	15.56
1	4	1.72	2.75	2.20	3.95	1	1	0.87	1.16	0.59	0.95
1	3	12.72	16.50	15.20	16.80	1	3	1.92	2.44	2.00	2.54
2	4	11.30	16.80	6.75	16.87	2	4	1.42	6.47	2.00	3.48
2	3	3.90	6.50	3.27	7.80	1	4	1.82	2.57	0.64	1.07
1	3	0.40	0.90	1.40	2.30	2	4	3.01	12.35	4.01	7.51
1	1	2.19	2.60	2.50	3.50	1	2	1.08	1.49	1.09	1.82
2	4	1.40	1.82	2.10	3.90	2	1	1.86	2.74	0.89	1.41
2	3	3.22	5.65	2.70	4.80	1	4	1.24	2.69	0.95	1.76
2	3	3.50	6.60	4.65	8.00	2	1	0.43	0.64	0.22	0.39
2	2	3.15	5.25	5.30	7.60	1	1	1.33	10.30	1.67	3.79
1	4	1.96	3.18	1.20	3.15	2	2	2.00	8.37	10.08	5.49
2	2	2.55	4.05	4.00	5.45	1	4	0.71	1.13	0.58	0.86
2	1	1.85	3.30	1.80	3.15	1	3	0.90	1.41	1.56	2.11
2	1	0.87	1.30	1.10	1.45	2	3	1.24	3.23	1.36	2.86
2	4	1.85	3.20	1.42	2.62	1	4	1.51	2.79	0.83	1.64
2	1	1.50	1.75	1.35	3.40	1	1	1.56	8.69	2.35	7.51
2	3	3.30	4.55	5.25	5.83	2	3	1.44	3.06	1.11	2.58
2	3	4.30	11.30	8.78	12.38	2	2	0.46	1.12	0.93	1.36
2	4	4.32	6.15	4.98	6.45	1	2	2.20	5.25	1.04	3.19
2	1	2.43	7.95	4.08	6.83	2	2	1.29	2.32	1.75	5.30
1	2	3.42	5.59	4.50	7.18	2	2	2.34	4.25	2.16	4.10
2	3	3.70	5.88	3.13	4.00	1	1	0.86	1.55	0.88	2.14
2	1	13.30	17.00	11.87	16.60	1	2	5.91	8.56	2.62	6.08
1	1	2.50	4.64	2.23	3.60	2	2	1.43	3.94	3.61	7.51

T2: as T1 but with one extra minute's exposure to the tone.

T3: as T2 but having been *instructed* to use the relaxation techniques of breathing and imagery. Note that there is no guarantee that subjects in groups T1 and T2 (and indeed the fourth group, T4, below) did not use the relaxation techniques. This was a deliberate element of the design, as other aspects of the experiment were aimed at studying motivation.

T4: this was a control group, in that the subjects experienced no exposure to the tone between (a) in the relaxation training and the final sensitivity measures.

Within the categories of migraine and tension, subjects were randomly assigned to the four treatment groups. Unfortunately, missing values reduced the complete data set from an intended balanced design to the group sizes shown in Table E.2. Table E.1 shows the raw data.

Table E.2 Group sizes for Study E.

	T1	T2	T3	T4
Migraine	11	11	12	11
Tension	14	11	16	12

The study is aimed at exploring whether the different treatments have different effects. In particular, the following contrasts on treatments are of interest:

C1:	1	−1	0	0
C2:	1	1	0	−2
C3:	0	0	1	−1

The first of these is the difference between groups T1 and T2; it thus explores whether an extra minute's exposure during the treatment phase has any effect. The second examines the difference between the control group, T4, and the average of the groups where no explicit instruction to use relaxation was given during the treatment phase (i.e. groups T1 and T2). The third examines the difference between T3, the treatment with the explicit relaxation instruction, and the control group, T4.

Since we are interested in the answer to the question addressed by each contrast in its own right, we shall carry out the test of each contrast at the 5% level. Furthermore, in testing each contrast we shall control for effects due to the other two contrasts. This means that any effects found can be unambiguously attributed to the contrast in question and cannot have arisen due to the relationships with the other contrasts. (That is, we will use the 'unique sum of squares' method.) Similarly, in examining these contrasts we wish to be certain

Headache

that any observed effects are not spurious ones induced from a real migraine/tension effect by the disproportionate group frequencies. We therefore eliminate effects due to the migraine/tension factor from our response when studying the above contrasts (see Section 2.10). Note that, since the migraine/tension factor only has two levels, the complete comparison for this factor consists of a single contrast:

$+1 \qquad -1$

Interest is centred on change from initial to final levels, so in this first analysis, rather than simply looking at the final U and DU scores (denoted U2 and DU2), we shall examine the difference scores between final and initial values. That is, if U1 and DU1 denote initial scores, we shall examine (U2 − U1) and (DU2 − DU1). Also, as noted above, primary interest is focused on general questions about sensitivity, and not on the U and DU changes separately (see Section 3.2). We therefore conducted a bivariate analysis of variance using the two variables (U2 − U1) and (DU2 − DU1).

The basic multivariate analysis of variance table is given in Table E.3 in a condensed form. As always when examining such significant tables we must first consider the interaction between the factors. If this is not significant then we can proceed to examine the main effects. In the case of this experiment we expected no interaction involving any treatment contrast and the migraine/tension factor. Thus our test of the interaction is merely a precautionary check on the overall assumption of no interaction. For this reason we analysed the complete set of interaction effects in a single test. If there were prior suspicions that some interaction contrasts were non-zero – that is, that some of the treatment effects took different values in the migraine and tension groups – then a different approach might be called for. In particular, one could then divide the interaction effect into its individual contrasts and consider each separately. For the way chosen in this present analysis, the multivariate test using the likelihood ratio criterion ((2) in Section 4.3) of the null hypothesis of no interactions (with three

Table E.3 The multivariate analysis of variance table for the (U2 − U1) and (DU2 − DU1) change scores.

Source	Between-groups d.f.	No. of variables	Wilks's statistic	P
Grand mean	1	2	.9255	.0320
Mig/Tension	1	2	.9851	.5123
Contrast 1	1	2	.9716	.2780
Contrast 2	1	2	.9393	.0616
Contrast 3	1	2	.9060	.0124
Interaction	3	2	.8337	.2122

(Error d.f. = 90)

160

contrasts and thus three degrees of freedom) gives a probability level of 0.212, certainly not significant at the 5% level.

We may now work up the table to the three treatment contrasts of particular interest. As described above, each of these three has been calculated in such a way as to answer the question: is this contrast necessary to explain between-group differences or are the other two adequate? Put another way, can all the between-group differences be summarized by only two contrasts, restricting the other to have a value of zero?

Recalling that we wish to test each of these three contrasts at the 5% level, we see from Table E.3 that only the third contrast is significant. It seems that the treatment has an effect on sensitivity change, but only when an explicit instruction to use relaxation techniques is given. Having said that, it is of interest to note that contrast C2,

| 1 | 1 | 0 | −2 |

almost achieves significance at the 5% level and might bear further investigation.

It is perhaps instructive to look at the application of a test of the complete treatment comparison. Assessing the three contrasts simultaneously in a single hypothesis matrix, testing the overall hypothesis that all contrasts are zero, yields a probability level of 0.063. If this were tested at the 5% level it would be non-significant – and one presumably would not go further to test the individual contrasts (except in an exploratory frame of mind). This, however, would be an unduly stringent test. An overall 5% level on a comparison comprising three contrasts would mean that, at the limit when the tests are independent, each was being tested at the 1.7% level (as calculated by the formula given in Section 5.4). Since in this study primary interest is focused on the three contrasts as individual questions, this seems a particularly severe level to adopt. Instead, we have adopted the 5% level for each contrast, which would yield a test level for the overall comparison of 14.3%, if they were independent. If this significance level is adopted then the 0.063 probability obtained for the overall comparison is comfortably significant. As always, the choice of significance levels must depend on deciding where one's interests lie.

Table E.3 shows the between-subjects degrees of freedom, the number of variables involved, the value of the multivariate test criterion and the associated probability level. The probability level of the likelihood ratio statistic is obtained after transforming the null distribution of the test criterion to an approximate F-distribution, as described in Section 4.3. For the sake of example we show, in Table E.4, the resulting approximating F-values and their associated degrees of freedom.

Finally, to round off the analysis, having explored the general question of sensitivity difference and established that there appears to be a difference between the third treatment group and the control, let us look at sequel questions and see if we can identify which, if not both, of the two variables

Headache

Table E.4 The approximating *F*-values and their associated degrees of freedom for the criterion values of Table E.3.

Source	Wilks's statistic	Approximating F-statistic	Associated d.f. (hyp/err)
Grand mean	.9255	3.581	2/89
Mig/Tension	.9851	.674	2/89
Contrast 1	.9716	1.299	2/89
Contrast 2	.9393	2.878	2/89
Contrast 3	.9060	4.616	2/89
Interaction	.8337	1.412	6/178

involved is responsible. To keep our overall level at 5% for this contrast it is appropriate to use a simultaneous test procedure. Using Gabriel's method (see Section 5.5) on this contrast alone, we compare the values of Wilks's lambda for the univariate tests with the required lambda value for the overall bivariate test 5% level. Note that, in the case of Wilks's lambda, values *smaller* than the percentage point mean that the null hypothesis should be rejected. In the present case, the 5% bivariate critical value for lambda is 0.9349. The observed univariate values are 0.9306 and 0.9439 for (U2 − U1) and (DU2 − DU1) respectively. Thus groups T3 and T4 seem to differ primarily on the (U2 − U1) change score.

E.2

Another way in which the analysis of these data might be undertaken is as follows. Instead of attempting to eliminate initial differences between groups by working with change scores, one could remove initial differences by using analysis of covariance. As it happens, this approach answers a subtly different question from that answered by the change scores analysis. However, so as not to interrupt this exposition, we shall leave the exploration of this difference until later.

Even having decided that one will use a covariance analysis approach, all is not settled. In this study we have two dependent variables (U2 and DU2) and two potential covariates (U1 and DU1) and these could be dealt with in at least two ways. One could adjust the U2 scores for U1 and the DU2 scores for DU1 or one could adjust U2 for both U1 and DU1 and DU2 for both U1 and DU1. One common way to interpret analysis of covariance is to regard the adjusted scores as the values the dependent variable 'would have taken' if all subjects had had the same score on the covariate. (We shall have more to say about interpreting analysis of covariance below.) The first suggested analysis above thus utilizes U2 values which would have obtained if all subjects had had identical U1 scores, and DU2 values which would have obtained if all subjects had had identical DU1 scores. The adjustment of U2 is not to the values which would have occurred if all subjects also scored identically on DU1. Similarly, the DU2 adjustment does not eliminate differences arising from different U1 scores. The second suggested method, however, also removes these other potential sources of difference. That is, the second method adjusts both the U2 and DU2 scores to the values they 'would have had' if the subject had all been identical on their initial U1 and DU1 scores. It seems to us that this second approach is much easier to interpret and this is the method we have adopted below.

A glance at the correlation matrix for these variables in Table E.5 reveals high correlations between the response variables (U2 and DU2) and the covariates (U1 and DU1), so that we may well expect the covariance adjustment to have a major impact on the estimates of effect sizes. The effect of these high correlations can be seen by comparing the unadjusted and adjusted standard errors of the effects in Table E.6. The latter are all much smaller than the former. In fact a likelihood ratio test of the null hypothesis that there is no effect of the two covariates on the bivariate response (taking the regression sums of products matrix and comparing it with the within-cells sums of products matrix, as described in Section 4.6) gives the F-statistic approximation to Wilks's lambda as $F(4,174) = 56.311$, which is significant beyond the 0.001 level. The regression coefficients are given in Table E.7.

Table E.5 Within-groups correlation matrix between U1, DU1, U2 and DU2.

	U1	DU1	U2	DU2
U1	1.000			
DU1	0.873	1.000		
U2	0.815	0.841	1.000	
DU2	0.805	0.873	0.869	1.000

Table E.6 Estimates of the three contrasts of interest before and after covariance adjustment, in both cases eliminating any induced effect from the migraine/tension factor.

	No adjustment		Adjustment	
	U2	DU2	U2	DU2
Estimated effect				
Contrast 1	−1.021	−0.592	−0.767	−0.204
Contrast 2	1.548	1.693	1.659	1.691
Contrast 3	2.194	2.429	1.513	1.534
Standard errors				
Contrast 1	0.913	1.234	0.476	0.599
Contrast 2	1.587	2.146	0.830	1.044
Contrast 3	0.880	1.190	0.461	0.580

Headache

Table E.7 Estimated regression coefficients for U2 regressed on U1 and DU1 and for DU2 regressed on U1 and DU1, both within treatment groups.

	U2	*DU2*
U1	0.3711	0.2664
DU1	0.3965	0.7034

Table E.8 Stepwise tests, first adding U1 and then DU1.

	P	*% additional variation explained*
Contribution of U1 alone		
U2	.0001	70.75
DU2	.0001	76.23
Bivariate P = .0001		
Additional contribution of DU1		
U2	.0036	2.71
DU2	.0905	0.76
Bivariate P = .0140		

It is possible that the high correlation between the covariates U1 and DU1 (0.873 from Table E.5) renders one of them superfluous (and, indeed, might have led to inaccurate estimated regression coefficients with large variances). This can be explored by studying the results of stepwise tests. Table E.8 shows the contributions resulting from using first U1 and then DU1. U1 makes a highly significant contribution to U2 and DU2, both separately and as a bivariate pair. Moreover, U1 accounts for about three-quarters of the variance in each of U2 and DU2. The effect of subsequently including DU1 as an additional covariate is shown in the lower half of Table E.8. Its contribution to U2 may be regarded as significant, that to DU2 will probably not be. The contribution to U2 and DU2 as a pair is significant at the 5% level.

Table E.9 shows the effect of first using DU1 as the sole covariate and then adding U1. The effects are all highly significant and the contribution of each covariate to the univariate variances of U2 and DU2 respectably high. In particular, the contribution of U1 to U2 and DU2 as a bivariate pair is highly significant. As a result, in what follows we have retained both covariates.

Table E.10 shows the bivariate analysis of variance results, having eliminated the covariate effects; Table E.11 shows the approximating F-values derived for the Wilks's lambda statistics in Table E.10. Again adopting separate 5% tests, we see that the interaction is not significant and that only the third contrast

0   0   1   −1

Table E.9 Stepwise tests, first adding DU1 and then U1.

	P	% additional variation
Contribution of DU1 alone		
U2	.0001	66.37
DU2	.0001	64.84
Bivariate $P = .0001$		
Additional contribution of U1		
U2	.0001	7.09
DU2	.0001	12.20
Bivariate $P = .0001$		

Table E.10 The multivariate analysis of covariance table for U2 and DU2, using U1 and DU1 as covariates.

Source	Between-subjects d.f.	No. of variables	Λ	P
Grand mean	1	2	.9399	.0674
Mig/Tension	1	2	.9803	.4216
Contrast 1	1	2	.9724	.2963
Contrast 2	1	2	.9457	.0877
Contrast 3	1	2	.8873	.0055
Interaction	3	2	.8729	.0628

Table E.11 The approximating F-values and their associated degrees of freedom for the criterion values of Table E.10.

Source	Wilks's statistic	Approximating F-statistic	Associated d.f. (hyp/err)
Grand mean	.9399	2.7839	2/87
Mig/Tension	.9803	0.8724	2/87
Contrast 1	.9724	1.2338	2/87
Contrast 2	.9457	2.5043	2/87
Contrast 3	.8873	5.5274	2/87
Interaction	.8729	2.0404	6/174

is significant. Thus only the treatment which includes the explicit instruction to use relaxation techniques shows an effect on the (adjusted) response variables U2 and DU2.

In parallel with the previous analysis of Section E.1, a simultaneous test procedure can be applied to the two response variables. The critical value of lambda in the overall test (yielding the 5% point in an approximating F-distribution with 2 and 87 degrees of freedom) is .9803. The values of lambda for the univariate tests on U2 and DU2 separately in this covariance analysis are .9007 and .9300 respectively. Recalling again that low lambda values indicate departure from the null hypothesis of no covariate effect, we see that this provides evidence that treatment groups T3 and T4 differ on both variables.

We have presented two analyses above, one in which any initial differences between groups were eliminated by measuring change from the initial values, and one in which any initial differences were eliminated by covariance analysis. The first merely eliminated U1 differences from U2 while the second also eliminated DU1 differences (with similar remarks applying to DU2). A priori one might have expected DU1 to have negligible effect on U2 once U1 had been eliminated, but we found this not to be the case.

In general, when one is presented with measurements taken at two time points there are several ways in which the analysis may be approached, but the most obvious are either to work with difference scores or to use the first occasion's measurements as covariates in analysing the second. These two approaches can yield different results – a fact which has led to some confusion in the past (and has been given the name of Lord's paradox). The reason that the results can differ is simply that the two approaches are asking different questions. This is most easily illustrated in the simple two-group comparison case; that is, we have two groups of subjects, each measured at two time points, and we wish to compare the changes the two groups experience. Then the difference-score approach enquires whether there is a difference in average change of the two populations. The covariance approach asks whether a member of group 1 is expected to change more than a member of group 2, given that they have the same initial value. It is this final condition which distinguishes the questions. In our case the two analyses were less comparable because DU1 was also used as covariate for U2 (as well as using U1 as a covariate) and similarly U1 as well as DU1 was used as a covariate for DU2. As it happened, however, our two analyses produced very similar results.

Smoking ——————————— Study F
styles

F.1

The data for this example are taken from a study of cigarette-smoking habits. The overall aim was to investigate differences in the ways in which people smoke cigarettes and the effects of different cigarettes upon the manner in which they are smoked. We are indebted to Dr S. Sutton, of the Addiction Research Unit, London, who provided these data and advised on their analysis.

From the pharmacological point of view, cigarettes can be characterized by their yield of tar, nicotine and carbon monoxide. These yields are measured by putting the cigarette in a machine which effectively sm kes it from beginning to end, and retains the smoke for chemical analysis. The machine-smoked yields of a cigarette cannot be thought of as directly comparable to the way any given individual smokes that cigarette, but they do provide a standardized way of assessing the yields of different cigarettes. These three measurements – tar yield, nicotine yield and carbon monoxide yield – are taken as the parameters which describe a cigarette.

An individual can vary the pharmacological effect a cigarette has on him or her in several ways. Most obviously the total volume of smoke passed into the mouth can be varied continuously. Secondly, the smoke extracted from the cigarette may be inhaled completely or partially. Finally, to some extent the individual can vary the richness of the smoke extracted from a cigarette by, for instance, drawing hard or gently on the cigarette, or by smoking the cigarette very heavily when it is near its butt end, where the concentration of tars will have built up. Further control of the yield of ventilated filter cigarettes can be achieved by, for instance, covering the ventilation holes in the filter as the cigarette is smoked. Many of these behaviours are unconscious actions, but none the less designed to produce the pleasing effect required.

It is not as yet possible to measure to what extent an individual inhales cigarette smoke. It is, however, possible to measure the amount of smoke he or she extracts from the cigarette into the mouth by means of a small recording attachment fixed to the butt end of a cigarette. This gives the measurement of total volume puffed smoke (TVPS). Measurement of the amount of nicotine and carbon monoxide (CO) actually taken in by a person can be made from a pharmacological analysis of blood samples. Firstly, blood plasma nicotine level (BPNL) and secondly, carboxyhaemoglobin measurements indicating the amount of CO absorbed by the blood stream (COHb) are usually made on blood samples given just after smoking a cigarette. Different individuals given identical cigarettes can vary these three measurements within certain physical limits as they wish, be it by conscious or unconscious means.

In the present study, 55 volunteers were studied in detail during and immediately after smoking a cigarette. The volume of inhaled cigarette smoke was measured in the manner described above. Two minutes after completion of the cigarette a blood sample was taken from the individual for the pharmacological analysis. Numerous other measures were made on the individuals in addition to these three smoking-style variables, for example age, sex, length of deprivation, and so forth. At an early stage of the analysis it was necessary to find out whether the smoking style of an individual would differ according to sex and age. The effect of age was anticipated to be non-linear, since it is usually the younger smokers who show the greatest variation in styles. These differences tend to disappear with age. Furthermore, there was a possibility that any age effects might have a structure among men different from that among women. The 55 smokers in the study were therefore classified by sex and by age group, the latter grouping con. sting of those in their twenties, those in their thirties, and so on. One individual was in his teens and was included with the twenty-year-olds, since by himself he would not provide enough stability to give information about this age group. For the same reason, two seventy-year-olds were included with the sixty-year-olds. The raw data for this study are given in Table F.9.

The first question of interest centres on the effects of age and sex upon the response variables. The between-subjects design consists of two factors, therefore, age at five levels and sex at two levels. It is a two-factor cross-classified design in which the main effects of sex and of age are expected to be important, but it will be necessary to check for the presence of an interaction effect between these two factors. The main effect of sex, having only one degree of freedom, is represented by the simple difference contrast

$$-1 \qquad 1$$

As mentioned above, the effect of age upon the responses may not be a linear one. If it were thought reasonable to assume that a linear relationship did hold, it would be possible to use the age measurement, without grouping into age categories, as a covariate in the analysis. However, the simplest way of checking for non-linear effects is to use age group as a categorical factor. For this factor we can define the complete comparison between the groups by a set of polynomial contrasts (Section 2.15). The first of these contrasts will represent roughly the linear effect of age. The second of these contrasts, the quadratic effect, should appear significantly different from zero if there are marked discrepancies from linearity in either of the extreme age groups. The third and fourth contrasts, the cubic and quartic, will allow the representation of more-complicated patterns.

168

Between them, this set of polynomial contrasts makes up the complete comparison and can represent any differences between the age groups.

The analysis uses the three response variables which describe smoking style: blood plasma nicotine level (BPNL), blood carboxyhaemoglobin (COHb), and volume of puffed smoke (TVPS). These three responses are treated as a profile of the smokers' style. No between-variables factor structure is imposed upon them, because prior linear transformations of them would not be meaningful. (Since these variables are not measured on the same scale, the average of them or arithmetic differences between them cannot be meaningfully interpreted.) We might note here, as a side issue, that one of the response variables – volume puffed – is in some sense logically prior to the other two responses: we might expect blood concentrations to depend on the amount of smoke extracted from the cigarette. This point is ignored in this analysis, but taken up in a later one.

The three response variables were not used in their raw form in this analysis, but were log-transformed at the outset. This transformation was chosen primarily because the relationships between the various measures in this study were thought to be multiplicative rather than additive. By taking logarithms of the variables such multiplicative models are transformed into additive models and these are the sort of relationships that analysis of variance is suited to investigate. The means of these three transformed variables for each cell of the classification, along with the number of observations in each cell, are given in Table F.1.

The multivariate analysis of variance for the data in Table F.1 is given in Table F.2. The method in which the 55 individuals for this study were sampled was to

Table F.1 The numbers of observations and the mean vectors of the log-transformed response variables: (a) blood plasma nicotine level (BPNL); (b) carboxyhaemoglobin (COHb); (c) total volume puffed smoke (TVPS) for each cell of the age by sex cross-classification.

	Age group				
	20s	30s	40s	50s	60s
Male					
Nos.	6	6	6	2	3
BPNL	12.90	12.44	12.59	12.29	11.90
COHb	11.29	10.95	11.29	11.02	10.85
TVPS	15.80	15.56	15.57	15.29	15.18
Female					
Nos.	11	7	9	3	2
BPNL	12.66	12.50	12.60	12.84	11.82
COHb	11.29	11.27	11.33	11.27	10.89
TVPS	14.45	15.36	15.40	15.58	15.50

Smoking styles

Table F.2 Analysis of variance for the data in Table F.1 ('unique' and 'sequential' sums of products calculation: see text).

Source	Sums of products			d.f.	Wilks's statistic
	TVPS	*BPNL*	*COHb*		
Within-cells	3.43	—	—		
E matrix	1.88	11.25	—	45	—
	0.57	3.48	4.69		
Constant **H** matrix	(not tested)			1	—
Sex	0.265	.	.		.881
H matrix	0.016	0.001	.	1	(sig .137)
	−0.219	−0.014	0.181		
Age (linear)	0.285	.	.		.846
H matrix	0.681	1.630	.	1	(sig .063)
	0.282	0.674	0.279		
Age (2nd, 3rd, 4th)	0.107	.	.		.830
H matrix	0.345	1.449	.	3	(sig .503)
	0.190	0.800	0.573		
Sex by age	0.561	.	.		.858
(linear)	0.342	0.209	.	1	(sig .083)
H matrix	0.039	0.024	0.003		
Sex by age	0.088	.	.		.910
(2nd, 3rd, 4th)	0.081	0.379	0.003	3	(sig .901)
H matrix	0.052	0.220	0.231		

recruit consecutive attenders at a 'smokers' clinic' into the study, with almost no clients refusing participation. This effectively makes the data a small-sample survey of a particular, narrowly defined population which we can call, for simplicity, 'reluctant smokers'. As such, the numbers of observations in the cells of the design are not due to experimental manipulation and choice, but rather are the natural occurrence of these types of individuals. It is therefore relevant to ask two types of question about their smoking behaviour. We might wish to assess the simple, observed effect of age upon the response variables: this option takes a *descriptive* approach (see Section 2.10). Alternatively, we might want to assess the effect of age after allowing for any possible effect due to sex which might distort the age group comparisons. It is this latter option, of assessing the *explanatory* effect of age (see Section 2.10) which is the one relevant to the researcher's question in this analysis. Similar considerations apply to sex differences. Thus the computed sums of products matrices shown in Table F.2

are made for each factor's main effect when controlling for the effect of the other; the interaction effects are calculated, as usual in a between-subjects design, controlling for both main effects.

This between-group breakdown of the variation in the response variables is shown in the leftmost column. The first entry is for the within-cells, or error, variation. The second entry is for the constant term of the between-subjects design, to be used in testing whether the grand mean across all subjects is zero. The entry for variation between the sexes follows, but for the main effect of age there are two entries in the table. The four degrees of freedom for the complete comparison of age groups have been partitioned into a component due to the first polynomial contrast (for linear trend across the groups) and a second component due to the higher-order polynomial contrasts (quadratic, cubic, quartic). We are primarily interested in the first contrast, which assesses the linear relationship of age with the response variables. The second comparison, which has three degrees of freedom, represents the pooling of any departures from this linear effect. The final two entries in the table are the corresponding interaction effects of sex with the linear trend due to age, and sex with the remaining age partial comparison. The first of these judges whether the linear relationship between age groups is the same for each sex; the second judges whether the remaining pattern of age effects shows a sex differential.

This is the between-subjects breakdown which would apply in any univariate analysis of variance. For the multivariate analysis of variance, instead of a sum of squares representing the variation due to any source we have the sums of products matrix as shown in the table. Thus the within-cells error is a three-by-three sums of products matrix with a column and a row for each of the three response variables. Were we to have performed three univariate analyses separately, the relevant sums of squares for each of the variables would be the entry in the diagonal of this matrix (see Section 3.1). In this study the zero point of the log-transformed response variables was of no particular relevance, so the entry for the constant term has been omitted from the table and will not be used in any test of hypothesis (see Section 2.9). For all other sources of variation a sums of products matrix is shown and in each case represents the hypothesis matrix to be used in the associated significance test.

The test statistic chosen for use was Wilks's criterion. This choice was made because there was no particular theory about the sort of departures from null hypothesis that might arise, so statistics which would be best against either extremes of noncentrality pattern (see Section 4.5) were avoided. Note that although this test statistic is quoted throughout the table, it is only when the hypothesis matrix has more than one degree of freedom that the choice actually matters – in the majority of tests in this analysis the one degree of freedom for hypothesis implies that all test statistics will give an equivalent result (see Section 4.4).

Thus beginning at the foot of Table F.2, to test the hypothesis that there is no interaction between the factor 'sex' and the higher-order polynomial differences between the age groups, the hypothesis matrix shown is compared with the within-cells error matrix at the head of the table. Furthermore, within the age comparison itself the higher-order terms are tested controlling for the linear term: this parallels the tests we wish to carry out on the main effect of age, where the aim is to see if a linear term alone is adequate to represent the data pattern, or whether the other terms add anything more to the explanation. A formal, and rather generous, significance level of .075 was chosen for the test of each of the null hypotheses, and it can be seen that Wilks's statistic does not achieve significance. (For the curious reader, we should say that this perhaps unusual significance level was chosen for didactic reasons in an unashamedly improper way.) Similarly, the hypothesis matrix for the linear relationship of age interacting with sex, when compared with the error sums of products matrix, gives no evidence for rejecting the null hypothesis.

Having tested the interactions and found no reason to reject the null hypotheses that these interactions are zero, we proceed to test the main effects. The effects of age are tested controlling for the effects of sex, for the reasons previously given. Within the age main effect the partitioned components are tested sequentially, as they were for the interactions. Thus, the test on the quadratic, cubic and quartic departures from linearity are tested for significance over and above the contribution of the linear trend itself. This test of the compound null hypothesis that all three contrasts are zero, controlling for the linear effect of age, is made in a way identical to that used for testing the interactions. That this test is not significant shows that the linear effect of age is an adequate model to account for the observed pattern of variation (see Section 2.10). Note that before the linear contrast is tested, we have only established that such a model for these effects is adequate, but it is possible that even that simple linear model is unnecessary. Perhaps even a less complicated model – that of no effects at all – is also adequate. However, the test of the hypothesis that the linear effect is null yields a significant value for Wilks's statistic and we conclude that there is indeed a linear effect of age shown in these data. The linear trend over age is a *minimal adequate model* for this main effect. The next highest entry in Table F.2 is the hypothesis matrix for testing the null hypothesis that there is no difference between the two sex categories, an hypothesis which is not rejected by the test of significance. As mentioned earlier, it is not relevant in this study to test the size of the constant term in the between-subjects design and the remaining hypothesis matrix is not tested.

We conclude therefore that the only significant effect in these data is the main effect of age, and that this effect is a linear trend of the response variables across the five age categories. Note that all these tests have been carried out simultaneously for the three response variables, since these were deemed to constitute the general smoking style of an individual. We have not sought in this analysis to establish whether the linear trend is present in all variables or, perhaps, in just one of them. What we have established is that a canonical variable (see Section 3.10) exists which does show a steady trend through age groups. The mean values of this canonical variable will be difficult to interpret in anything but an abstract sense. However, the simple pattern of these means in the four groups is given in Table F.4. This is the type of answer we obtain to asking a general question (see Section 3.2) about groups' profiles. If there had been more than one degree of freedom in the hypothesis we would have obtained several canonical variables, and perhaps several that were significant. However, a one-degree-of-freedom hypothesis requires only a single canonical variable (see Section 3.10) to represent it. From the coefficients given in Table F.3, which define the weights on the variables used in constructing this canonical variable, we see that the predominant variable is the blood plasma nicotine level of the smoker. In this analysis we do not proceed to ask detailed follow-up questions (type (c) in Section 4.1) about this point, because the primary question – do smoking styles differ – has been answered. The secondary purpose in the original study centred on a more formally structured question, to which we turn in the next analysis.

Table F.3 Coefficients used to construct the canonical variable associated with the linear trend in the age effect in the analysis of Table F.2.

TVPS	BPNL	COHb
−0.445	−0.665	−0.189

Table F.4 Mean value in each age group of the canonical variable and the associated estimate of the age (linear) effect (see Table F.2).

	20s	30s	40s	50s	60s
Mean values	−.660	−.330	0.0	.330	.660

Estimated effect: 0.330

F.2

In this study of styles of smoking, the primary interest was, as stated above, in whether differences existed between age groups and between the sexes. Although only a simple trend across age groups was found significant (and that at the 7.5% level), we shall none the less proceed to the secondary question of the study. Interest here focuses on the mechanisms by which this age difference might be produced. There are two separate questions that the researcher wished to ask: firstly, were any differences that were found simply due to people selecting different types of cigarette; secondly, were the achieved pharmacological effects of smoking these cigarettes mediated entirely by the most obvious method of manipulation at the smokers' disposal, that of varying the total volume of puffed smoke? We shall take each of these questions in turn and see that the answers come in each case from a multivariate analysis of covariance.

The first question concerns the type of cigarette chosen by the smoker. For the purposes of this study, the type of cigarette was characterized by the three pharmacological parameters already described: measures of machine-smoked yield of tar, nicotine and carbon monoxide. If different subgroups of people achieve their different responses simply by choosing different cigarettes, then by controlling for the type of cigarette that is being smoked differences between subgroups should disappear. On the other hand, the responses there may still be directly affected by age or sex, even if we have standardized them for differences in chosen type of cigarette. This question is therefore answered by an analysis of covariance, in which differences between the groups defined by the between-subject factors are assessed after controlling for the effect of the covariates nicotine yield, carbon monoxide yield and tar yield (see Section 3.8).

Table F.5 displays this analysis of covariance. As usual in any covariate analysis, the first entry is for the within-cells adjusted error, which refers to error variation after controlling for the effect of the covariates. It is usually termed the sums of products about regression. The second entry relates to variation due to the regression and indicates the strength of the effect of the covariates upon the responses. The remaining entries in the table give the usual breakdown for the between-subjects design: firstly, the constant or grand mean term; secondly, the main effect of the 'sex' factor; and finally, the last two entries are a partitioning of the main effect of the 'age' factor. This partitioning of the effects of age, as was described in the previous analysis, separates the linear effect of age upon the responses (with one degree of freedom) from the remaining, more complex patterns of differences between the groups (three degrees of freedom, representing the second, third and fourth polynomials). Associated with each entry is a sums of products matrix for the three response variables, BPNL, COHb, and TVPS

After the within-groups error matrix, denoted $\mathbf{E}$ in the text (see Section 4.2), here adjusted for the regression, the second entry is the matrix of sums of products due to regression. This we treat as any other hypothesis matrix (those denoted $\mathbf{H}$ in the text) when formulating a test of significance. Note that the regression hypothesis matrix has three degrees of freedom associated with it because it is formed from regressions on three predictor variables, namely the covariates giving the three cigarette-machine-smoked yields (the within-cells error degrees of freedom have correspondingly been reduced by three). The test of the null hypothesis that the regression coefficients are all zero, i.e. the covariates have no effect upon the responses, is therefore formulated in the

Table F.5 Analysis of covariance for the data in Table F.1 (main effects only), using machine-smoked cigarette yields as covariates ('unique' sums of products calculations).

Source	Sums of products			d.f.	Wilks's statistic
	TVPS	BPNL	COHb		
Within-cells (adjusted) E matrix	2.266 1.800 0.478	. 11.218 3.427	. . 3.894	42	—
Regression H matrix	1.163 0.083 0.095	. 0.025 0.053	. . 0.800	3	.498 (sig .001)
Constant	(not tested)			1	—
Sex H matrix	0.288 0.022 −0.190	. 0.002 −0.015	. . 0.126	1	.844 (sig .076)
Age (linear) H matrix	0.523 0.858 0.236	. 1.406 0.387	. . 0.106	1	.788 (sig .022)
Age (2nd, 3rd, 4th) H matrix	0.188 0.264 0.039	. 1.214 0.495	. . 0.284	3	.820 (sig .511)

usual manner by computing the eigenvalues of the matrix ratio $\mathbf{H}\cdot\mathbf{E}^{-1}$ (see Section 4.3). It is again a difficult decision in this analysis to choose between the various test statistics which can be derived from these eigenvalues, since no prior expectations exist about the nature of the alternative hypotheses. For demonstration purposes Wilks's criterion was selected as a measure intermediate between those which are suitable for concentrated noncentrality patterns and those which are suited for diffuse patterns. Selecting a significance level of .10, we see that the multivariate test for the simultaneous null hypothesis on all regression coefficients is highly significant. The nature of this regression relationship between the responses and the covariates is examined in detail at a later point.

As mentioned earlier, tests on the constant term, or grand mean, are not relevant in this analysis because no particular importance attaches to the average response levels in the data. We are concerned here solely with differences between groups, and the hypothesis matrix for the constant term has not been shown in Table F.5. The next entry is a sums of products matrix which assesses

175

the strength of the 'sex' contrast of the between-groups design after controlling for the other main effect (of 'age') and the effects of the covariates. We can see that Wilks's statistic for testing the hypothesis that there is no difference between the two sex groups in any of the response variables easily achieves our chosen significance level. It is worth noting that the size of this effect, as indicated by the sums of squares for each response variable in the diagonal of the matrix, is slightly larger than the effect when measured without controlling for the covariates (see Table F.2). The bottom two entries in Table F.5 relate to the main effect for age, controlling for the effects of sex and the three covariates. Within the main effect for age, the two tests are sequential, in that the bottom entry for the sums of products due to the higher-order polynomials are computed controlling for the linear age effect, whereas this linear age effect is not controlled for the higher-order polynomials, exactly as in the previous analysis. We can see from the test of the higher-order polynomial contrasts that they contribute nothing above and beyond the linear contrast which itself is highly significant. We thus conclude that the effect of age can be represented by a simple linear trend across the age groups.

It is perhaps worth pointing out that both the linear effect of age and the effect of sex on the responses are a little more clearly revealed in this analysis of covariance than they were in the simple multivariate analysis of variance of the preceding section (see Table F.2). This is probably a result of the marginally reduced within-cells error in the covariance analysis. As in ordinary univariate analysis, whenever we succeed in controlling for a significant source of error in the data – here the significant relationship of the covariates with the responses – we thereby increase the accuracy of the experiment. This is true whether or not the covariates are distributed so as to introduce bias into the between-group comparisons. It is simply due to the increased variation they induce in the response variables, which is removed into the regression hypothesis matrix.

In Table F.5, the multivariate test of the regression hypothesis matrix rejected the compound null hypothesis that all regression coefficients were zero. Table F.6 presents the estimated values of these regression coefficients. Each response variable – that is, puffed volume of smoke, blood plasma nicotine level and carboxyhaemoglobin – was regressed on the three covariates simultaneously – tar yield, nicotine yield and CO yield – and the squared multiple correlation is computed for each of these three regression analyses. These are the usual computations that are found in any univariate multiple regression. From the squared multiple correlations, which give the proportion of variance explained by the three yields, we can see that blood plasma nicotine level is not predictable from them. The third measure, COHb, is predictable to a certain, small extent from the tar yield and the CO yield of the cigarette. The predictability of puffed volume of smoke is moderately high and comes jointly from the tar yield and nicotine yield of the cigarette. It would be possible to summarize these relationships by using the canonical variables representation of the variation due

176

Table F.6 Within-cells regression analysis associated with the covariates for the analysis in Table F.5.

	Response variables		
	TVPS	*BPNL*	*COHb*
Regression coefficients (standardized) for covariate:			
Tar yield	−0.97	−0.15	−0.59
Nic. yield	0.63	0.10	−0.07
CO yield	0.01	0.09	0.55
Squared multiple correlation			
	.339	.003	.170

to regression and the regression weights which best predict them. This is not, however, our concern in this study.

In Table F.5, as in any covariance analysis, the regressions have been calculated within each cell of the between-subjects design, that is, for each age and sex group, and the results pooled to produce a single regression for each of the response variables. Implicit in this procedure is the assumption that the regression relationship is the same for all groups, an assumption which has not been tested in this analysis.

We conclude from the analysis that there is a strong effect of cigarette type, as represented by its yields, on the response variables and in addition to this some further effect due to the age and sex of the individual.

F.3

We turn finally to the second question asked above, which concerned how the differences in smoking styles are primarily achieved. It was mentioned earlier that the total volume of puffed smoke is in some sense a variable which is logically prior to the other two responses. In the chronological sequence of events it precedes the other two and, moreover, it is the direct and most obvious means that the smoker has of manipulating his or her pharmacological state. Having established that age and sex affect the trio of response variables, even when we allow for the pharmacological differences in the cigarettes that people choose, we now ask whether these effects are mediated entirely by the volume of smoke they puff. Thus we wish to inspect the data for differences in the two pharmacological blood measures after we have allowed for any differences that may be induced by the volume of smoke which has been drawn from the cigarette. If simple volume is the complete mediator of the demographic variables' effect upon blood concentrations, no such systematic differences between the various groups of people should be found. Conversely, if different ages and different sexes can find some other means of inducing different concentrations, these will show through as age and sex effects upon the two blood concentration measures even after controlling for the volume of smoke puffed.

Note that this single question about the relationship between the responses is all that is being asked. We are not hunting through all possible component hypotheses of the overall hypothesis ('no response differences') to single out those which can be accepted and those which are rejected. Neither are we interested in a sequential test of a series of hypotheses, such as the Roy–Bargmann step-down tests provide – our choice of hypothesis is restricted and clearly defined. The techniques for looking at such sub-tests (described in Chapter 5) are not required here. The analysis required is a single multivariate analysis of covariance again, as in the preceding analysis, but this time one of the response variables (volume of puffed smoke) is instead to be regarded temporarily as another covariate. The remaining two variables (BPNL and COHb level) constitute a bivariate response whose between-group variation we wish to analyse.

Table F.7 shows this analysis. It is a parallel to the analysis given in Table F.5, but the sums of products matrices here are for just two response variables, the pharmacological blood measures. The between-subjects design is exactly that of the analysis in Table F.5, with hypothesis matrices for the main effects of age and

Table F.7 Re-analysis of the data in Table F.1, using total volume of puffed smoke (TVPS) as a covariate in addition to the three cigarette yields.

Source	Sums of products		d.f.	Wilks's statistic
	BPNL	COHb		
Within-cells error **E**	9.788 3.047	. 3.794	41	—
Regression matrix **H**	1.465 0.432	. 0.901	4	.684 (sig .046)
Constant matrix **H**	(not tested)		1	—
Sex matrix **H**	0.131 0.160	. 0.194	1	.951 (sig .369)
Age (linear) matrix **H**	0.303 0.086	. 0.025	1	.970 (sig .542)
Age (2nd, 3rd, 4th) matrix **H**	0.907 0.440	. 0.276	3	.888 (sig .560)

Table F.8 Within-cells regression analysis associated with the covariates for the analysis in Table F.7.

| | Response variables | |
	BPNL	COHb
Regression coefficients (standardized) for covariate:		
TVPS	0.44	0.18
Tar yield	0.27	0.42
Nic. yield	−0.18	−0.18
CO yield	0.08	0.54
Squared multiple correlation		
	.13	.19

sex and for the overall constant. The hypothesis matrix for the variation due to the regression relationship is present as before, but is now based on four covariates, the original three cigarette yield measures and in addition the measure of volume of puffed smoke. Note that it has associated with it four degrees of freedom as a result. Interpreting this table in an exactly analogous way to Table F.5, we see that there are now no significant effects for age or for sex. Only the regression on the covariates is significant (again, the constant term is not of interest to us here).

Table F.8 gives the regression coefficients relating to the hypothesis matrix of Table F.7 and the squared multiple correlation coefficients from the regressions. Neither of the two responses is well predicted from the covariates, but the usual univariate significance tests on these regression coefficients show us (a) that volume of puffed smoke is a significant predictor of blood nicotine level, and (b) that only the CO yield of the cigarette predicts the COHb measure.

Putting the results from these two covariance analyses together, we can present the information in diagrammatic form, as is shown in Fig. F.1. The volume of puffed smoke does indeed mediate the effects of age and sex upon the

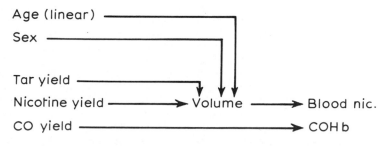

Figure F.1 Results from Tables F.6 and F.8.

179

other two response measures. It is the only variable which predicts the nicotine blood level and is itself determined by the tar and nicotine yields of the cigarettes. Note that tar yield has a negative and nicotine yield a positive effect on the total volume of smoke puffed. By contrast, the COHb level is determined directly by the CO yield of the cigarette and by this variable alone. Subsequent analyses, not reported here, showed these conclusions to be unchanged by the introduction of other control factors – specifically length of the cigarette and of deprivation – which, it turned out, had no additional effects on response.

Table F.9 Raw data for Study F.

	Sex	Age	TarY	NicY	COY	TVPS	BPNL	COHb
1	2	3	180.00	15.00	183.00	3758.00	123.00	70.00
2	2	3	160.00	10.00	177.00	4017.99	311.00	100.00
3	2	4	160.00	10.00	177.00	4739.00	242.00	121.00
4	2	2	180.00	13.00	177.00	6827.79	474.00	91.00
5	2	4	190.00	12.00	183.00	3074.19	375.00	103.00
6	2	5	180.00	15.00	183.00	6716.99	449.00	87.00
7	2	4	170.00	13.00	165.00	4065.79	419.00	88.00
8	1	6	160.00	11.00	133.00	4405.59	185.00	53.00
9	2	6	190.00	14.00	196.00	4575.40	33.00	53.00
10	2	6	180.00	15.00	183.00	6343.99	564.00	54.00
11	1	4	90.00	5.00	144.00	6788.99	256.00	91.00
12	2	4	180.00	13.00	188.00	3539.79	242.00	97.00
13	1	2	180.00	13.00	177.00	7273.79	312.00	42.00
14	2	4	80.00	7.00	113.00	7446.19	179.00	61.00
15	1	3	190.00	14.00	196.00	7369.39	456.00	85.00
16	2	3	190.00	13.00	157.00	4150.79	232.00	55.00
17	2	5	190.00	15.00	190.00	5688.99	389.00	84.00
18	1	4	160.00	11.00	133.00	6341.59	429.00	57.00
19	2	5	180.00	13.00	188.00	5194.79	309.00	66.00
20	2	3	190.00	12.00	183.00	5005.19	269.00	69.00
21	2	2	150.00	7.00	201.00	4270.20	274.00	108.00
22	1	4	180.00	15.00	183.00	6164.99	274.00	84.00
23	2	4	120.00	13.00	115.00	6410.79	157.00	50.00
24	1	2	180.00	15.00	183.00	7169.99	348.00	95.00
25	2	2	130.00	9.00	168.00	4538.39	384.00	119.00
26	2	2	180.00	14.00	180.00	4521.39	274.00	72.00
27	2	2	150.00	11.00	156.00	5931.60	179.00	59.00
28	1	5	190.00	14.00	196.00	5400.39	306.00	66.00
29	2	2	190.00	12.00	183.00	3590.19	260.00	46.00
30	2	3	90.00	9.00	123.00	8502.39	346.00	86.00
31	1	3	180.00	15.00	183.00	6828.99	106.00	28.00
32	1	4	190.00	14.00	196.00	6256.39	468.00	96.00
33	1	2	180.00	13.00	177.00	8045.80	597.00	155.00
34	2	2	180.00	15.00	183.00	5251.99	233.00	84.00
35	2	2	190.00	14.00	196.00	5849.39	304.00	94.00

Table F.9 (continued)

	Sex	Age	TarY	NicY	COY	TVPS	BPNL	COHb
36	2	4	90.00	9.00	123.00	7225.39	448.00	109.00
37	1	4	90.00	5.00	144.00	5761.99	182.00	109.00
38	2	3	180.00	15.00	183.00	7766.99	527.00	100.00
39	1	5	190.00	14.00	196.00	3541.39	155.00	56.00
40	1	2	190.00	12.00	183.00	6004.19	347.00	79.00
41	1	6	190.00	13.00	157.00	3559.79	74.00	34.00
42	2	2	120.00	6.00	191.00	5812.59	471.00	86.00
43	1	3	180.00	15.00	183.00	6469.00	260.00	61.00
44	1	3	190.00	14.00	196.00	5735.39	213.00	43.00
45	2	4	190.00	14.00	196.00	4650.39	346.00	71.00
46	1	2	180.00	15.00	183.00	7114.00	304.00	73.00
47	1	4	180.00	15.00	183.00	3869.99	256.00	56.00
48	1	6	90.00	8.00	110.00	3894.79	233.00	76.00
49	2	3	190.00	12.00	183.00	2366.20	227.00	80.00
50	1	2	180.00	15.00	183.00	8424.99	607.00	75.00
51	2	2	180.00	13.00	171.00	6232.79	464.00	81.00
52	2	4	190.00	14.00	196.00	4863.39	469.00	77.00
53	1	3	180.00	13.00	177.00	2780.80	209.00	46.00
54	2	2	190.00	12.00	183.00	4724.19	314.00	71.00
55	1	3	150.00	11.00	156.00	6638.59	456.00	120.00

Alzheimer's disease ——————— Study G

G.1

We are indebted to Professor Raymond Levy, Miss Adrienne Little and Dr Paz Chuaqui for permission to use the data on which this example is based.

The senile dementias affect approximately one in twenty of the population aged over 65. Traditionally, it has been assumed that these conditions involve an inevitable and progressive deterioration in all aspects of intellect, self-care, and personality. However, recent work suggests that one senile dementia (Alzheimer's disease) involves pathological changes in the central cholinergic system which could produce the intellectual deficits found in this condition, especially the deterioration in memory. Now, it might be possible to remedy this cholinergic dysfunction and thereby reverse, halt or slow down the memory impairment by long-term dietary enrichment with lecithin, which is a precursor of choline. The first step in investigating this hypothesis is to find a suitable measuring instrument.

There are many problems in monitoring the effects of treatments in such diseases. It is likely that any effects will be small and slow so that tests of memory must be sensitive to subtle but clinically significant changes. Most available tests are vulnerable to many sources of error. One important source of error relates to the need (by definition) to administer tests on several occasions in order to monitor change. It is reasonable to expect subjects' test performance to improve over assessments without any underlying improvement in cognitive ability, simply because of a practice effect, i.e. a positive carryover from one test occasion to the next. The standard way to avoid this is to use parallel forms of test, these being equivalent but different tests. (Usually the tests involve words, and the different tests involve different but equivalent sets of words.) Clearly, while these control for carryover due to repeated exposure to the same items, they do not control for carryover due to repeated exposure to the test situation.

However, we can make these apparent deficiencies work to our advantage. After all, the aim of the treatment is to alter the performance of the memory function; and carryover effects are measures of memory functions. Thus we can compare the extent of practice effect of the repeated forms test with that of the parallel forms test and use this comparison as an index of the cognitive effect of the treatments in Alzheimer's disease. This will be an index which can be used in addition to the basic time trends observed in test performance.

The data as presented were two randomly assigned groups of subjects, 26 taking a placebo, and 25 taking lecithin. Three of the lecithin group had missing data items and have been omitted from the analysis presented here. The fact that three in this group but none in the placebo group had missing values arouses suspicion. Are the lecithin group more likely to drop out than the placebo group, and does this distort the results? Unfortunately, the answer to the first of these questions appears to be yes. The subjects were less able to take the lecithin than the placebo without developing unpleasant side-effects. This is a weakness in the design. The answer to the second question is unclear. We must assume that the reasons for dropping out do not affect the response test scores. In this experiment, preliminary t-tests of the initial test scores (see below) on the two groups gave $t = 1.050$ ($P = 0.299$) and $t = 0.505$ ($P = 0.616$) for the two test types respectively, neither of which is significant.

Every subject received either the placebo or lecithin for a six-month period, during which time the cognitive tests were carried out at the start, at one month, at two months, at four months, and at six months. Each test involved the subject being given a list of words and then being immediately tested to see how many he or she could recall. Each subject received both types of test: type 1, the repeated form test, with the same words each time; type 2, the parallel form test, using different but equivalent sets of words each time. The raw data are given in Table G.1. Mean scores are plotted in Fig. G.1.

The major question of interest is whether memory effect differs between the two treatment groups. We can explore this by studying the pattern of change in scores over time. We therefore begin by transforming the repeated tests over time to polynomial components. That is, in place of the five measures at times 0, 1, 2, 4, and 6 months, we calculated the overall mean, the linear trend, the quadratic trend, and so on. These provide a clearer description of chronological change. Note that in carrying out this trend transformation it is necessary to take into account the fact that the measurements are not taken at equally spaced intervals. This means that non-standard coefficients have to be used for the polynomials rather than the standard ones discussed in Section 2.5. For example, the coefficients for the normalized linear component were:

$$-0.5398 \qquad -0.3322 \qquad -0.1246 \qquad 0.2907 \qquad 0.7058$$

Note that the interval between the first two is half the interval between the last two, reflecting the spacing of the time points. These coefficients, and those for the other polynomial components, were generated with a computer program, as part of a multivariate analysis of variance package.

This transformation leads to polynomial descriptions of both test types. Thus each subject has overall mean scores for types 1 and 2, overall linear trend scores for types 1 and 2, overall quadratic scores for types 1 and 2, and so on through cubic and quartic (4th power) sources.

We will be interested in knowing whether or not the two test types yield

Table G.1 The Alzheimer data. Column (1) shows group (1 = placebo, 2 = lecithin), columns (2) to (6) are the five time points for test type 1, and columns (7) to (11) are the five time points for test type 2.

(1)	(2)	(3)	(4)	(5)	(6)	(7)	(8)	(9)	(10)	(11)
1	20	19	20	20	18	15	9	13	15	14
1	14	15	16	9	6	11	14	11	12	6
1	7	5	8	8	5	2	3	6	10	7
1	6	10	9	10	10	5	4	6	2	1
1	9	7	9	4	6	6	5	8	6	3
1	9	10	9	11	11	5	11	7	4	8
1	7	3	7	6	3	3	1	8	4	5
1	18	20	20	23	21	22	11	21	18	18
1	6	10	10	13	14	13	14	10	14	10
1	10	15	15	15	14	3	7	12	10	6
1	5	9	7	3	12	8	6	5	6	7
1	11	11	8	10	9	4	12	7	6	4
1	10	2	9	3	2	10	4	9	6	0
1	17	12	14	15	13	14	13	15	12	13
1	16	15	13	7	9	11	6	19	3	6
1	7	10	4	10	5	10	3	10	1	1
1	5	0	5	0	0	3	0	7	0	0
1	16	7	7	6	10	18	14	10	7	11
1	5	6	9	5	6	9	2	10	8	3
1	2	1	1	2	2	6	1	6	10	4
1	7	11	7	5	11	3	15	5	13	6
1	9	16	17	10	6	3	3	7	8	5
1	2	5	6	7	6	3	6	9	9	3
1	7	3	5	5	5	5	0	8	3	2
1	19	13	19	17	17	18	10	11	11	9
1	7	5	8	8	6	8	1	11	10	3
2	9	11	14	11	14	9	8	14	6	6
2	6	7	9	12	16	8	5	6	4	6
2	13	18	14	20	14	7	19	10	22	7
2	9	10	9	8	7	6	11	9	7	6
2	6	7	4	5	4	1	1	0	5	7
2	11	11	5	10	12	5	3	10	10	4
2	7	10	11	8	5	3	9	6	12	4
2	8	18	19	15	14	18	22	22	17	17
2	3	3	3	1	3	2	0	1	2	0
2	4	10	9	17	10	9	7	6	20	11
2	11	10	5	15	16	14	17	7	19	14
2	1	3	2	2	5	6	6	11	9	7
2	6	7	7	6	7	5	9	3	7	3
2	0	3	2	0	0	0	0	1	0	0
2	18	19	15	17	20	20	14	13	14	17
2	15	15	15	14	12	9	15	11	11	9
2	14	11	8	10	8	15	6	11	8	8
2	6	6	5	5	8	3	5	8	1	6
2	10	10	6	10	9	6	4	8	10	10
2	4	6	6	4	2	4	3	3	3	4
2	4	13	9	8	7	11	4	11	8	5
2	14	7	8	10	6	6	8	10	12	5

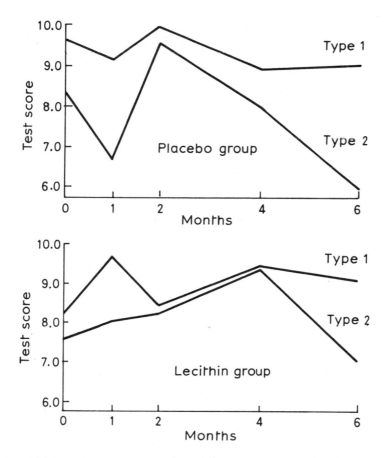

Figure G.1 Mean scores in Alzheimer's disease study.

different chronological patterns. The answer to this question sheds light on whether the repeated forms tests (type 1) are more sensitive to treatment effects than parallel forms tests (type 2). Furthermore, if there is no difference in time pattern between the two test types then we can pool the results to yield more-reliable overall estimates of time changes.

To answer this question of differences in time patterns it is convenient to make yet another transformation. For simplicity let us first focus attention on just the linear components of the time changes of the two test types. Now, as far as the linear component goes, the question of difference of time pattern is addressed by the *difference* between the linear component for test type 1 and that for test type 2. Similarly for the quadratic, cubic, and quartic components. If there is no difference between the linear components we can fruitfully average them – and the same for the other components. Thus the new transformation takes the two sets of four polynomial components and calculates one set of four differences and one set of four averages (or, equivalently, sums). Note that we have not mentioned the missing polynomial components, the means of the two test types over time, in this description. The means tell us nothing about chronological *patterns*, but about levels. The difference between the mean scores for the two test types thus addresses a different question, one to which we will return below.

The transformations described above are derived from the *between-variables* design of type by time, as described in Section 3.5. The between-variables factor 'type' has two levels (repeated and parallel) and its comparison, having only one degree of freedom, is represented by the single contrast

$-1 \qquad +1$

and the overall level is represented by the linear combination

$+1 \qquad +1$

When these are taken in conjunction with the polynomial trend representation of the 'time' comparison, the ten measures in the between-variables design are transformed into the sets of variables described. These sets are presented in Table G.2 (the extension to more-complicated designs, with multiple degrees of freedom in the factors, should be evident from this table).

Consider firstly the most complex of these sets of variables, the difference scores resulting from these transformations: when we average over groups, looking at the grand mean of the data on these difference scores, we will be studying the 'type' by 'time' interaction; if, however, we *compare* the different groups using these scores we will be studying the effect of group on this 'type' by 'time' interaction. One might regard this as a sort of 'group by type by time' interaction (Section 3.5).

When we use the four averages (or sums) to compare groups, we are addressing a 'groups' by 'time' interaction. Using the same average scores and looking at the mean over the two groups simply yields the overall time pattern, the comparison for the 'time' main effect.

If we now turn to the remaining polynomial component, the means over time in each of the two test types, we find: using the difference of these means to make the between-groups comparison gives the 'group' by 'type' interaction; using the average of the means to make this between-groups comparison gives the group main effect; and the mean over the two groups of the difference between test type 1 and test type 2 gives the comparison for the test type main effect.

Before we proceed to calculate any of these effects or interactions we must examine the data. There appear to be no extreme outliers, but there is some evidence of skewness to the right. Since detailed questioning of the researchers revealed that the raw scores were in fact counts out of 30, we carried out a logistic transformation. On this new scale the data seemed satisfactorily symmetric.

Table G.3 summarizes the effects. Note that in this table the first four rows

Table G.2 The between-variables and between-group design for the Alzheimer's disease memory test experiment.

Derived variables in the between-variables design	Between-group design	
	Grand mean of group 1 and group 2:	Comparison of group 1 with group 2:
Polynomial trends: (average over types)		
Mean	Overall response level (Grand mean)	Group comparison
Linear		
Quadratic	Time	Group × time
Cubic	comparison	
Quartic		
Polynomial trends: (difference between types)		
Mean	Type comparison	Group × type
Linear		
Quadratic	Type × time	Group × type × time
Cubic		
Quartic		

involve univariate tests and thus the test statistics follow the F-distribution exactly, provided the assumptions and the null hypothesis are true. The other rows are multivariate tests and have been obtained using an F-distribution approximation to statistic (2) of Section 4.3.

We shall use 5% significance levels in this analysis. As always, we must start to interpret these results by looking at the highest-order interaction. In this case it is the effect of group on the 'type by time' interaction and it is clearly non-significant, meaning that any relationship between the type of test and the pattern of scores over time is the same in each group (or, more rigorously, we have no reason for supposing them not to be the same).

Since there is no effect of group on the 'type by time' interaction variables, we can collapse the two groups when we explore this 'type by time' interaction. Moving up to the second line from the bottom of Table G.3, we find that 'type by time' is highly significant. That is, it appears that the two test types evolve differently over time. We shall return to this point, but first let us develop the results in a different direction.

Alzheimer's disease

Table G.3. Multivariate analysis of variance results (in summary form) for the data in Table G.1. The hypothesis and error sums-of-products matrices have not been shown. The number of degrees of freedom associated with the error matrix is 46.

Source	d.f.	No. vars.	Wilks's	F (exact)	P
Grand mean	1	1	.365	80.494	.000
Group	1	1	.999	0.011	.919
Type	1	1	.762	14.397	.001
Group × type	1	1	.994	0.283	.597
				F (approx.)	P
Time	1	4	.753	3.535	.014
Group × time	1	4	.841	2.032	.107
Type × time	1	4	.674	5.051	.002
Group × type × time	1	4	.957	0.487	.745

Recall that the primary question referred to a comparison of the two groups, one using lecithin and the other a placebo. It is these treatment group differences we shall detail first. We have established that the two groups do not affect the 'time by type' *relationship* in different ways (i.e. the differences in the polynomial trends for test type 1 and test type 2), but they might still affect the general time evolution pattern in different ways; and group 1 might still exhibit a different 'type' effect from group 2. We can explore the time evolution question for the two groups by averaging over the 'type' factor to produce the group by time row in Table G.2. (This collapsing is legitimate because the between-variables part of the design is balanced.) This averaged 'time' effect is not significantly different between groups at the 5% level. As with the highest-order interaction, we could now collapse the groups and explore the pattern of score evolution over time (a significant result in row 5 of Table G.3), but since this is not a primary stated interest we shall not do so here.

An exactly parallel argument in exploring the 'type' comparison leads us to examine row 4 (group by type – not significant) and row 3 (type – not of primary interest although significant). Finally, in row 2 we see that not even the overall levels of the group scores seem to differ. No effects involving differences between groups, neither main nor interaction effects, seem to exist.

Returning now to the significant type by time interaction, recall that a secondary question was whether the two types evolved differently over time. The multivariate test on the four type by time interaction variables showed that this indeed appears to be the case. A natural sequel question is then to characterize this different evolution.

Table G.4 shows the sequence of univariate tests applied to the four polynomial components of the type difference scores. That is, the top row refers to the linear components of the type evolution over time, the second row refers to the difference in quadratic components, and so on.

We must decide at what level each of the four tests might be conducted. The tests will not be independent, so working out a precise overall significance will be difficult. (Note that step-down tests would be independent. However, as we have explained in Section 3.8, under normal circumstances step-down tests are not appropriate for answering these sorts of questions.) We could apply Gabriel's test, as described in Section 5.5, but this seems overly conservative since it will protect us against incorrect rejection of the null hypothesis on any possible linear combination of the four variables. For simplicity we shall merely

Table G.4 Univariate sequence of tests for the difference in each polynomial component mean between the two test types (the 'type by time' interaction variables). The number of degrees of freedom associated with the error term is 46.

Variable	d.f.	F	P
Linear	1	0.6624	.420
Quadratic	1	7.2489	.010
Cubic	1	6.2158	.016
Quartic	1	6.2884	.016

take an upper bound by calculating at what level each variable should be tested if we knew that they were independent and wished an overall 5% level of protection. As we have discussed in Section 5.2 such a value is found by setting

$$(1 - \alpha)^4 = 1 - 0.05$$

Calculation gives $\alpha = 0.012$ in this formula.

Referring to Table G.4 we must begin with the highest-order term, i.e. the quartic in row 4. Using the above significance level this is found to be non-significant. So also is the cubic component. The quadratic, however, has a probability level of 0.010 and is significant. Thus the two types of test seem to differ in their quadratic shape – and perhaps in their linear slope. However, having found the quadratic difference it is not of interest to go on to test this: our testing sequence stops when we reject some null hypothesis, and so we declare a significant effect of at least quadratic order. Hence, a different pattern of evolution over time for the two test types has been established.

It is perhaps worth recording at this point that contrasts other than the polynomial trends are often used for analysing repeated measures designs. For instance, by extending the principle of unequal spacings it is possible to go further and make a pre-specified rescaling transformation of the time-point labellings. Thus for processes which we might expect to reach an asymptote (or limit) as time progressed we could relabel the time points as exponentially increasing or decreasing intervals. This would make, for instance, simple linear trends a more appropriate description of the time change pattern. In this example such a relabelling did not appear to be appropriate.

Psychophysiological — **Study H** responses

H.1

We are grateful to Dr John Brown of the Institute of Psychiatry for permission to use the data below, which were collected during a study investigating how psychophysiological responses to a disturbing stimulus were affected by the degree of predictability of the stimulus.

The between-subjects design was a simple two-by-two factorial. The first factor, 'condition', was an *experimental* factor (meaning that the experimenter could control to which level the subjects were assigned), in contrast to a *classification* factor with subjects randomly allocated to the two groups. The two levels of this 'condition' factor were named 'predictable' and 'unpredictable', as explained below. The second factor, termed 'preference', had levels 'preferred predictable' or 'preferred unpredictable', these labels indicating which of the condition levels the subject would have preferred to be assigned to, given a choice. Since the experimenter had no control over which level the subject preferred, this second factor is a *classification* factor.

Of the 79 original volunteers, 2 declined to complete the experiment and 17 were eliminated from the analysis due to recording faults. In such a situation it is very important to check that no selection bias in the elimination is occurring. For instance, it is conceivable that those who were allocated to the condition which they did not prefer experienced some kind of stress which affected the effectiveness of the contact of the electrodes measuring the physiological processes. Figure H.1 illustrates how the 60 remaining subjects were distributed across the four groups, with the bracketed figures showing the distribution of the

	Condition	
	Predictable	Unpredictable
Preference — Predictable	20 (6)	17 (4)
Unpredictable	10 (2)	13 (5)

Figure H.1 Distributions of the 60 analysed subjects across cells. The distribution of the 17 subjects who dropped out is shown in brackets.

Psychophysiological responses

17 who dropped out. As can be seen, there is a broadly equitable distribution between the four cells. Ideally one should carry out an examination to see if the drop-outs differ from those remaining in each cell. In the present case, however, the numbers are so small that the difference would have to be very large to have any hope of obtaining significance. A cursory examination of the pre-experiment measures that were available showed no large differences, so it was decided to proceed under the explicit assumption that the drop-outs had not introduced any bias.

Each subject was faced with a panel of ten lights, numbered 1 to 10. The lights came on one at a time, each remaining illuminated for five seconds. When light number 7 came on a 106 dB tone sounded. This tone was the disturbing stimulus. In the 'predictable' condition, light number 7 was preceded by a consecutive sequence of two, three, or four lights. Thus one of the light sequences (5, 6, 7), (4, 5, 6, 7) or (3, 4, 5, 6, 7) occurred. This means that in the 'predictable' condition, as soon as the light sequence had started the subject knew whether he had 10, 15 or 20 seconds to wait before light number 7 came on and the tone sounded. However, the order of presentation of the three possible light sequences was randomized (in a way explained below) so that, until the sequence had begun, the subject could not predict the waiting time. The reason for this rather complicated arrangement was to make the predictable and unpredictable conditions more directly comparable.

In the 'unpredictable' condition the exposure also consisted of a set of two, three, or four lights followed by light number 7 and the tone (the subjects did not know that only three sequence lengths were possible). The difference between this and the predictable condition was that here the lights were not constrained to be consecutive, but could occur in any order. (For example, the sequence 4, 6, 3, 5, 7 might occur.) Thus in this condition the commencement of a light sequence did not enable a subject to predict whether he had 10, 15, or 20 seconds to wait.

Each subject experienced ten exposures to light/tone stimuli, each exposure consisting of a light sequence (of 10, 15, or 20 seconds) followed by the tone. Of the ten exposures, the first, fourth, seventh, and tenth for both condition groups were of length 15 seconds before light 7 came on. The remaining exposures were of length 10 or 20 seconds, three of each in a random order. The object of this was to make the unpredictable waiting time less predictable. (Obviously if all pre-tone sequences had been of length 15 seconds then, no matter what the light order, the waiting time would have been predictable.) Rather than randomizing the ten exposures totally, the regular pattern of the 15-second exposures was adopted to make any time effect more comparable for the 'predictable' and 'unpredictable' groups – in both groups the first, fourth, seventh and tenth were of length 15 seconds. Only data from these four exposures were analysed. In the description of the analyses we term these four exposures 'trials' and label them from 1 to 4.

192

Each trial fell naturally into eight 5-second time bands: two preceding the commencement of the light sequence, three while the three consecutive lights preceding number 7 were on, one for light number 7, and finally two following it. Heart rate was computed during each of these time bands using a cardiotach. This measures the time interval between heart beats, converts it into a rate by taking the reciprocal, and the experimenter then averaged these rates over the 5-second interval to give a final heart rate figure for that interval.

The average of the rates in the first two of the eight time bands yielded a baseline measure. The baseline measure resulting from the first trial was used as a covariate in the analysis; the baselines from the three succeeding trials were ignored. Within each trial, the remaining six time bands fell naturally into two groups of three. The first three were termed the 'anticipation' group and the last three 'recovery'. The word 'epoch' is used to describe each of the three time bands in each of these groups. The analysis described below simply looks at the three anticipation epochs.

Summarizing, the between-variables design is a four (trial) by three (epoch) factorial design, the between-subjects design is a two (condition) by two (preference) design, and there is one covariate (baseline). Figure H.2 illustrates this. The data are given in Table H.1.

Perhaps it is worth remarking that an alternative possible design would have been to give each subject both predictable and unpredictable stimuli, as a way of further eliminating between-subject variation. There were two reasons for not adopting this approach. One involves theoretical psychological reasons, which we need not go into here, and the other is the simple risk of carry-over effects. Previous similar studies have found such effects.

At this point we recommend that the reader casts an eye over the raw data in Table H.1 to see if anything is striking. An example of the sort of thing that may be noticed is the abundance of low values in the data – an overall average baseline heart rate of 68.36. This is quite low, especially for people in an experimental situation. It is perhaps explained by the fact that most of the subjects were aged under 25 and some were athletically inclined, and so, fit. There is also another point which can be seen in the raw data, and one which is very important. We shall return to it below, but deliberately omit describing it now in order to permit the reader to exercise his or her own skills.

Since the three anticipation epochs occur in each of four trials there are twelve dependent variables in all. We could thus simply carry out a twelve-variable, two-by-two multivariate analysis of covariance. However, whether or not differences are discovered, one is almost certain to want to explore matters further. For example, if differences between groups are found, we would like to know if this is due to differences in behaviour over the three epochs, or due to differences in behaviour over trials. Furthermore, by perspicacious choice of alternative variables – that is by choice of some transformation applied to the twelve original variables – we might be able to simplify the description. For

Table H.1 Raw data for Study H.

Column (1) is condition (1 = predictable). Column (2) is preference (1 = predictable). Columns (3) to (14) are the 12 dependent variable scores in the following order (where Ti is trial i and Ej is epoch j): T1E1, T2E1, T3E1, T4E1; T1E2, T2E2, T3E2, T4E2; T1E3, T2E3, T3E3, T4E3. Column (15) is the covariate.

(1)	(2)	(3)	(4)	(5)	(6)	(7)	(8)	(9)	(10)	(11)	(12)	(13)	(14)	(15)
1	1	61.14	67.94	66.63	66.44	63.81	68.17	64.33	65.22	62.45	68.73	63.72	63.30	62.12
1	2	71.78	68.78	74.31	72.16	69.95	69.67	72.11	70.42	69.72	71.54	72.06	70.09	75.39
1	1	69.02	66.30	63.11	65.45	68.83	64.42	64.33	68.97	69.67	71.41	63.53	69.16	64.87
1	1	65.73	63.48	64.14	63.81	67.80	67.52	69.63	66.58	68.17	68.31	72.02	67.70	64.11
1	1	66.06	70.75	64.38	65.08	65.22	62.50	63.44	64.52	64.47	66.20	64.05	63.86	66.95
1	1	81.25	79.38	76.52	75.48	79.70	81.77	84.77	73.23	79.70	84.39	77.08	73.89	72.20
1	1	63.81	60.44	61.33	69.02	64.84	62.55	62.13	66.44	72.30	65.08	58.66	66.11	70.77
1	1	68.13	72.48	66.58	70.42	68.92	68.82	65.50	65.92	69.25	66.77	63.02	67.75	66.77
1	1	70.66	69.39	72.06	71.13	71.36	67.89	70.47	72.39	69.77	68.83	71.74	70.94	69.90
1	1	100.09	105.39	90.30	99.25	96.91	107.27	98.03	97.14	103.38	106.94	98.69	99.11	102.67
1	1	94.09	84.67	93.02	92.55	93.02	89.64	83.08	92.64	96.39	87.34	94.09	85.98	90.11
1	2	71.97	70.94	73.09	71.69	73.94	72.67	72.30	71.97	74.36	78.11	72.63	70.89	72.65
1	2	67.28	68.22	69.81	70.42	67.52	67.28	71.08	69.06	67.00	67.09	65.36	69.48	67.37
1	2	68.41	61.05	65.88	65.64	73.23	66.20	66.06	65.36	72.25	62.03	60.20	71.03	65.57
1	2	57.34	58.61	57.58	54.77	56.31	60.30	58.28	66.06	60.81	63.53	60.95	69.95	56.17
1	1	64.89	59.13	58.09	59.03	71.45	58.80	60.63	59.13	81.39	63.06	65.55	64.05	65.31
1	2	64.05	67.75	69.48	62.27	74.55	68.50	75.06	66.30	70.09	68.41	72.02	63.11	68.57
1	2	65.73	69.48	66.30	65.92	69.95	67.75	67.42	70.61	68.78	67.56	68.08	60.86	68.24
1	1	73.33	73.23	78.77	75.39	74.64	72.81	76.98	77.92	72.58	73.56	72.77	77.78	74.33
1	1	61.89	59.82	58.61	59.88	60.20	60.44	60.72	57.30	63.06	62.45	59.03	61.52	57.70
1	1	53.36	57.20	52.23	50.13	53.03	53.83	52.65	52.98	49.52	54.77	52.38	51.25	58.93
1	1	63.02	61.42	60.72	62.27	64.65	62.17	63.86	62.64	68.08	58.75	61.42	61.52	66.37
1	2	75.58	70.09	72.16	61.23	77.69	74.27	69.30	66.02	65.36	64.56	72.81	70.14	72.67
1	1	63.67	67.61	64.52	71.55	64.94	62.13	67.70	64.14	70.94	69.58	67.84	66.63	59.76
1	1	58.98	57.63	58.98	56.55	58.47	57.77	61.23	56.88	58.94	60.25	57.67	57.48	56.36
								59.79		61.70	57.01	58.75	61.00	59.40

G1	G2													
1	1	60.48	58.00	56.45	58.56	62.50	57.48	59.36	59.08	63.67	58.84	58.28	57.72	59.47
1	2	82.52	80.69	85.14	85.75	82.66	81.25	78.25	84.30	83.36	82.75	83.41	79.33	83.76
1	2	83.36	87.02	85.23	80.73	82.84	85.84	83.17	80.41	83.31	84.63	81.85	81.44	83.73
2	1	62.23	60.82	64.04	60.47	63.16	59.82	70.20	60.35	63.22	61.29	56.95	60.52	61.99
2	1	61.28	61.32	62.07	63.06	60.63	61.80	61.89	64.19	61.89	62.92	62.64	64.60	61.42
2	1	77.73	70.05	82.05	75.25	77.73	72.34	77.88	75.43	80.31	75.30	75.25	67.66	77.92
2	2	72.20	70.33	74.59	70.98	70.28	67.98	73.80	71.59	70.89	69.39	73.23	71.13	68.59
2	1	54.95	54.86	56.03	55.52	56.64	54.81	53.36	55.23	61.89	53.78	53.41	55.98	54.93
2	2	65.73	62.78	54.20	61.75	69.58	66.25	53.83	57.30	68.59	63.86	55.61	60.30	64.49
2	1	62.69	64.00	69.63	72.44	58.09	60.72	66.34	66.25	65.22	60.30	66.25	67.61	68.38
2	2	64.23	66.11	65.31	65.36	62.83	65.36	62.83	69.06	63.95	65.88	63.44	65.45	65.05
2	2	78.63	70.80	74.55	72.67	73.05	71.03	72.48	73.19	72.72	71.83	71.55	71.92	79.79
2	2	67.56	63.72	64.70	57.11	62.22	63.25	67.05	61.52	61.70	59.41	63.39	59.97	65.80
2	1	84.06	87.72	89.22	79.38	78.25	76.95	85.84	77.13	77.55	76.75	80.97	81.77	84.53
2	2	70.80	68.08	68.41	70.52	69.63	72.67	67.00	65.64	69.30	68.78	69.39	64.09	67.09
2	1	70.56	62.31	63.58	64.61	68.36	63.11	64.00	64.52	72.67	62.55	66.11	66.86	69.11
2	1	68.27	67.89	67.28	66.63	66.53	65.22	67.23	68.78	67.19	65.83	65.69	68.22	68.73
2	1	82.75	70.14	77.78	79.33	85.98	75.34	74.45	79.51	83.97	73.80	73.89	77.08	83.76
2	2	101.31	92.03	92.92	90.48	103.00	96.30	90.44	90.48	108.53	98.55	90.34	90.91	94.87
2	2	72.34	71.73	81.02	72.25	75.91	69.11	76.14	74.92	79.19	69.39	77.41	77.41	74.50
2	2	60.72	59.13	62.59	62.78	58.89	63.63	58.56	59.45	59.17	61.19	59.08	59.97	61.35
2	2	64.19	63.63	71.64	75.67	67.14	65.83	68.03	70.38	59.92	64.70	67.00	66.11	65.41
2	1	69.16	78.67	74.69	69.44	64.23	67.61	67.89	71.73	65.64	72.25	65.83	70.05	66.81
2	1	66.44	67.84	68.17	68.45	69.34	67.52	67.70	68.92	68.59	66.53	69.20	67.52	66.13
2	1	56.08	56.13	57.48	58.32	56.83	58.94	57.72	59.13	61.75	62.50	57.06	58.61	57.91
2	2	72.86	71.97	71.27	79.52	73.47	72.44	71.92	66.72	72.39	74.13	70.84	76.47	70.28
2	2	86.13	92.97	85.70	90.58	87.77	85.33	85.19	89.13	88.70	88.05	86.03	93.39	84.20
2	1	69.86	69.11	78.11	75.25	67.38	67.66	71.88	70.80	65.50	67.33	69.44	66.72	70.58
2	2	61.09	64.14	58.66	58.56	61.05	64.61	59.50	60.06	58.56	59.69	58.84	55.89	61.89
2	1	56.36	54.58	56.41	59.88	55.75	54.72	55.75	58.94	60.67	57.81	54.91	57.16	57.36
2	2	61.89	60.34	57.81	59.41	57.77	60.72	58.00	57.34	60.77	59.31	58.75	57.34	57.34
2	2	53.69	56.59	53.27	56.36	55.14	56.82	55.05	56.55	55.84	56.50	54.86	56.03	54.74
2	2	57.76	60.20	61.70	58.98	55.23	56.08	64.94	56.45	57.02	55.61	63.20	54.20	59.40

Psychophysiological responses

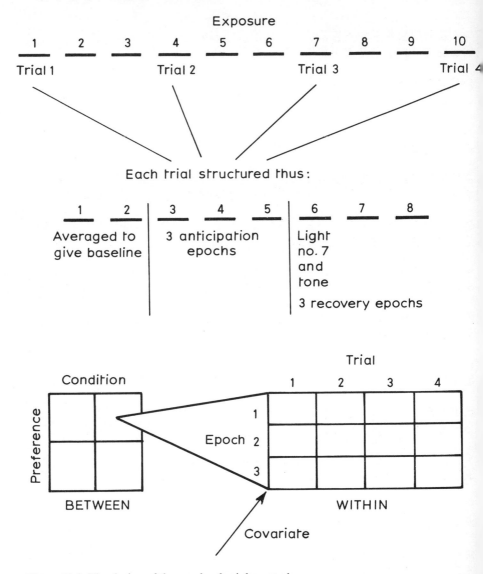

Figure H.2 The design of the psychophysiology study.

example, we might find that a parsimonious description of change between trials was given by a simple linear trend and that any more complicated pattern of differences could be attributed to error variation. The questions being asked, then, are not general questions, but questions involving the structure of the between-variables design.

For these reasons we transform the twelve dependent variables into an alternative twelve which focus better on the above questions. We stress that the two sets of variables are equivalent in the sense that each is describing the same thing: any data point defined in terms of one set has a corresponding point in the other, and it is possible to transform either set to the other. The point is simply that the new set is a more convenient description and one which might allow us to ignore some of the variables and so lead to a simpler description of the data. The new twelve variables fall into four blocks:

(1) A grand mean over all variables.
(2) A set of two variables representing the comparison between epochs (computed by averaging each epoch over the four trials and then calculating contrasts between these averages). The chosen contrasts were the linear and quadratic trends since these seem to provide a convenient description for our purposes.
(3) A set of three variables representing the comparison between trials, analogous to 2. Again polynomial components were used, namely linear, quadratic, and cubic trends.
(4) A set of six variables representing interactions between the epoch and trial factors. These new variables describe differences between trials on the patterns of change over epochs 1, 2, and 3.

The transformation matrix to produce these new variables will be quite complicated. One could calculate it manually, but to avoid the tedium we used the computer program to calculate it for us. Most multivariate analysis packages permit this, using symbolic descriptions of the contrasts required. Some programs transform all the input variables whereas in this example we did not want the covariate included in any transformation. In such a case it is necessary to carry out a dummy run just using the twelve dependent variables in order to generate the transformation matrix, and then perform a second run, reading in the transformation matrix enlarged by an extra row and column for the covariate. The matrix is given in Table H.2.

We tackled the analysis by dividing the twelve variables into the four blocks shown above and analysing each block separately. This immediately presents a question since the sets of variables fall into a natural order (from 1 to 4) should we, at each stage, covary out the preceding sets?

Set 2 describes epoch effects and set 3 trial effects. Covarying epoch out of the trial analysis means that we will be examining between-experimental-group differences using only those parts of trial not correlated with epoch. This might be the question the experimenter wishes to answer, but we have assumed he or she is more concerned with the question: are there between-group differences measured on the trial variables regardless of any correlations with epoch variables?

Psychophysiological responses

Table H.2 Transformation matrix: original variables index the columns; derived variables index the rows.

	T1E1	T2E1	T3E1	T4E1	T1E2	T2E2
M1	0.289	0.289	0.289	0.289	0.289	0.289
E1	-0.354	-0.354	-0.354	-0.354	0.0	0.0
E2	0.204	0.204	0.204	0.204	-0.408	-0.408
T1	-0.387	-0.129	0.129	0.387	-0.387	-0.129
T2	0.289	-0.289	-0.289	0.289	0.289	-0.289
T3	-0.129	0.387	-0.387	0.129	-0.129	0.387
X1	0.474	0.158	-0.158	-0.474	0.0	0.0
X2	-0.354	0.354	0.354	-0.354	0.0	0.0
X3	0.158	-0.474	0.474	-0.158	0.0	0.0
X4	-0.274	-0.091	0.091	0.274	0.548	0.183
X5	0.204	-0.204	-0.204	0.204	-0.408	0.408
X6	-0.091	0.274	-0.274	0.091	0.183	-0.548

	T3E2	T4E2	T1E3	T2E3	T3E3	T4E3
M1	0.289	0.289	0.289	0.289	0.289	0.289
E1	0.0	0.0	0.354	0.354	0.354	0.354
E2	-0.408	-0.408	0.204	0.204	0.204	0.204
T1	-0.129	0.387	-0.387	-0.129	0.129	0.387
T2	-0.289	-0.289	0.289	0.289	-0.289	0.289
T3	0.387	-0.129	-0.129	0.387	-0.387	0.129
X1	0.0	0.0	-0.474	-0.158	0.158	0.474
X2	0.0	0.0	0.354	-0.354	-0.354	0.354
X3	0.0	0.0	-0.158	0.474	0.091	0.158
X4	-0.183	-0.548	-0.274	-0.091	-0.204	0.274
X5	0.408	-0.408	0.204	0.204	-0.274	0.204
X6	0.548	-0.183	-0.091	0.274	-0.274	0.091

If trial effects were found which disappeared when epoch variables were controlled for, then one might attribute the apparent trial effects to the epoch effects, but this would be a difficult interpretation. It seems to us much clearer to examine the variables sets separately and ask two distinct questions: are there between-group effects on the epoch variables and are there between-group effects on the trial variables? This extends to include all four groups of variables, so we have carried out four separate analyses of covariance, one on each of variable sets 1 to 4.

As far as set 4 goes, we are asking the question: are there differences between the experimental groups in terms of the epoch by trial interaction variables? Simply that, and *not* eliminating any such differences arising due to, or explainable by, correlations between these interaction variables and the lower-order epoch and trial variables.

Table H.3 Analysis of covariance for the variables grouped in blocks 1 to 4. The baseline is used as the covariate. The hypothesis and error sums of products matrices have not been shown; the number of degrees of freedom associated with the error matrix is 55.

Source	d.f. for F	F	P	Univariate sig. level	
1 Grand mean (1 variable):					
Pref.	1/55	.015	.904		
Cond.	1/55	.563	.456		
Pref. cond.	1/55	.120	.731		

Effect	d.f. for F	Approx. F	P	Linear	Quad.
2 Epoch (2 variables):					
P E	2/54	1.698	.193	.160	.233
C E	2/54	4.347	.018	.012	.159
P C E	2/54	1.906	.159	.234	.127
3 Trial (3 variables):					
P T	3/53	.733	.538		
C T	3/53	.653	.585		
P C T	3/53	1.060	.374		
4 Epoch by trial (6 variables):					
P E T	6/50	2.065	.074		
C E T	6/50	.529	.784		
P C E T	6/50	1.770	.124		

P: Preference. C: Condition. E: Epoch. T: Trial.

The comparison for the preference (between-group) factor was not of any theoretical interest in this study but, as mentioned earlier, was included as a classification factor simply to control for any effects it might induce on the response variables. The researcher's main interest lay in the explanatory factor of experimental condition – the comparison of predictable and unpredictable stimuli, controlling for any preference effects.

In Table H.3 the basic results are shown in condensed form. Within each block of variables, the sums of products hypothesis matrices for tests of the between-groups comparisons have been calculated controlling for the preceding effects (referred to as a 'sequential' presentation). The matrices themselves and Wilks's statistics have not been shown, simply to make the table more easily assimilable.

Using a 5% significance level for each multivariate test, a glance at the probability levels shows that the only significant result is for the comparison of conditions on the epoch variables. This effect is significant at the 5% level, indicating that the two groups vary in different ways as the sequence of three lights comes on. We can decompose this (bivariate) test into its linear and quadratic components and examine each of these separately using univariate

Psychophysiological responses

Table H.4 Mean scores for the three epochs, averaged over trials, for the two condition groups separately, (a) adjusted for the covariate, (b) unadjusted.

Condition	Epoch		
	1	2	3
(a)			
Predictable	68.00	68.62	68.97
Unpredictable	68.54	67.36	67.33
(b)			
Predictable	68.21	68.63	68.97
Unpredictable	68.33	67.36	67.44

tests. The significance level chosen for each of these tests was .025, calculated (from the formula given in Section 5.4) so that

$$1 - (1 - .025)^2 = .05$$

We find that only the linear component of the epoch time pattern demonstrates a significant difference between the two condition groups. (The use of step-down tests is not appropriate here, as was discussed in Section 5.4.)

To summarize, it seems that there are no differences in trial time pattern between the four groups, but there are differences in epoch time pattern between the two condition groups and these are adequately described by a straightforward linear trend. The means for the three raw epoch variables in the two condition groups are shown in Table H.4 and Fig. H.3.

H.2

Throughout all of the above analysis the average of the first two (trial 1) baseline measures was used as a covariate. There are two points relating to this that we should consider:

(a) Is there advantage to be gained by this covariance analysis? Perhaps the reduction in error variance due to using a covariate is not sufficient to counterbalance the reduction in the number of degrees of freedom due to estimating the regression parameters.

(b) Are the regression slopes homogeneous across groups? This is a standard analysis of covariance prerequisite and we must ask if it is reasonable in the present context. Needless to say, these points should have been considered at the start, and we have left them to this stage for purposes of exposition.

Examination of the within-cells correlations between each of the twelve dependent variables and the covariate reveals a highly significant correlation of

0.956 between the covariate and the grand mean dependent variable (that comprising block 1) and no other significant correlations. The raw regression coefficient of the grand mean dependent variable on the covariate is 3.242 and highly significant ($P < .001$). What does all this mean?

First, the highly significant regression slope of the grand mean variable is not surprising on theoretical grounds. A subject with a high (low) baseline might reasonably be expected to have high (low) scores on the other variables.

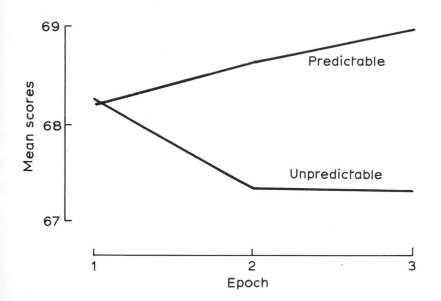

Figure H.3 Mean scores averaged over trials.

Second, the other dependent variables are contrasts on repeated measures, and so any constant baseline is subtracted out. The non-significant regression coefficients here thus mean, for example, that the linear trend for a subject is not affected by his original baseline level. That is, the situation appears to be as in Fig. H.4(a) rather than H.4(b). Since the regression coefficients of the dependent variables (in blocks 2, 3 and 4) are non-significant there seems little justification for using the covariate.

Table H.5 shows the results of the analysis without the covariate. Very little difference is discernible between this and Table H.3 (except for the grand mean variable, which is not of primary interest). This seems to suggest that, as far as patterns of change over time are concerned, the covariate is irrelevant. We shall not pursue these considerations here, nor the testing of the homogeneity of the covariance slopes, but instead move on to a more useful analysis.

Psychophysiological responses

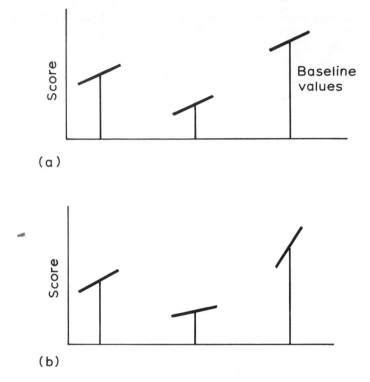

(a)

(b)

Figure H.4 (a) The slope for each subject is the same for each baseline value. (b) The slope depends on baseline level.

H.3

Let us now return to the cryptic suggestion made earlier, that the reader should examine the data rather more closely than we have up to now.

The data reveal a large number of scores in the 50s and 60s and a very few in the 100s. This can be seen just by glancing at the values, without any particular care being taken, and it suggests that it might be worthwhile examining the distribution more closely. Recall that multivariate analysis of variance, as with the univariate case, requires certain conditions to be satisfied. For example, for the tests to be valid the raw data must derive from populations with normal distributions. If we have a moderately large sample size, and if other assumptions are satisfied (such as equal variances), then the analysis is robust to non-normality (within reason). However, in our case the sample is not large.

Figure H.5 shows histograms of the first six raw variables. Positive skewness is evident in all the variables. (The values under the figures give skewness calculated from the raw data.) Thus, it seems that in order for the analysis to be

Table H.5 As Table H.3 but without using a covariate in the analysis. The hypothesis and error sums of products matrices have not been shown; the number of degrees of freedom associated with the error matrix is 56.

Source	d.f. for F	F	P	Univariate sig. level	
1 Grand mean (1 variable):					
Pref.	1/56	0.644	0.426		
Cond.	1/56	0.216	0.644		
Pref. cond.	1/56	3.102	0.084		

Effect	d.f. for F	Approx. F	P	Linear	Quad.
2 Epoch (2 variables):					
P E	2/55	1.738	0.186	.131	.290
C E	2/55	4.137	0.021	.013	.174
P C E	2/55	1.258	0.292	.319	.127
3 Trial (3 variables):					
P T	3/54	0.644	0.590		
C T	3/54	0.686	0.564		
P C T	3/54	1.372	0.261		
4 Epoch by trial (6 variables):					
P E T	6/51	2.210	0.057		
C E T	6/51	0.525	0.787		
P C E T	6/51	2.146	0.064		

P: Preference. C: Condition. E: Epoch. T: Trial.

justified, we must transform the data so that they more closely resemble a normal distribution. Of course, ideally these assessments of skew and normality should be made within each cell of the analysis, so that we are sure they are not induced by experimental effects. One way of doing this rather crudely in small samples is to subtract the appropriate cell mean from each observation before drawing up the overall distribution. In the present example this procedure gives roughly comparable results.

To many people, the idea of transforming data seems intuitively unreasonable. They argue that they want to work with the thing actually measured, not some arbitrary transformation of it. Sometimes it might be possible to maintain this position, in which case, since the data are not of the correct form to be eligible for classical analysis of variance, classical analysis of variance cannot be applied. A nonparametric technique – making no normality assumptions – must be used.

In other cases the argument is not so convincing. In the present case, heart rate was calculated by measuring between-beat time intervals, taking the reciprocal, and averaging over a five-second period. So, where are the raw data? The

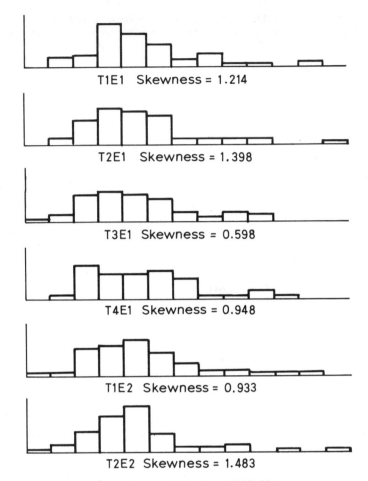

T1E1 Skewness = 1.214

T2E1 Skewness = 1.398

T3E1 Skewness = 0.598

T4E1 Skewness = 0.948

T1E2 Skewness = 0.933

T2E2 Skewness = 1.483

Figure H.5 Histograms of the first six variables from Table H.1.

numbers we have been analysing have already been subjected to two transformations (reciprocal and average). In such a case it does not seem very
convincing to argue that heart rate is more pertinent than log (heart rate) or
square-root (heart rate). Since one of these, or some other transformation, may
yield distributions conforming to the assumptions of the parametric analysis, it is
sensible to carry them out and use such an analysis. Recall that, when the
assumptions are justified, a parametric approach generally has greater power
than a nonparametric approach.

The reader might like to look at this issue another way. Our aim is to compare
the groups using some kind of average or representative values for the groups.
And for skewed distributions the arithmetic mean is not a very good rep-

resentative value. Transforming the data and then calculating the arithmetic mean is equivalent to using some other kind of summary measure on the raw data.

Ideally, what is needed at this point is some kind of within-group multivariate transformation to normality of all 12 dependent variables simultaneously. This, however, is impracticable, and even if straightforward methods were available it would be unreliable to perform such a transformation based on 60 data points measured on 12 variables. For this reason we content ourselves with carrying out 12 separate univariate transformations. This leads to approximately normally distributed marginals – but this does *not* imply that the multivariate distribution is normal. After all, the marginal distribution of a variable takes no account of its relationship to any other variable. We must still make the explicit assumption that the relationship between the variables is such that, given normal marginals, there is a multivariate normal distribution. By virtue of this assumption, the transformed variables we will obtain in blocks 1 to 4 will also be normal and hence may legitimately be subjected to multivariate analysis of variance.

For any particular variable within some particular group, our aim is ideally to discover the distribution of that variable so that we can derive a transformation which will change the distribution to normality. This implies that we should consider using different transformations for each variable within each group (if their basic distributions are different then different transformations may be needed to bring them to normality). But if we use different transformations then we are treating the groups and variables differently and this would defeat the object of the study, which is primarily to compare the condition groups.

Permitting different transformations in fact leads to an analysis which answers a different question. If, for example, the skewness of a raw distribution was proportional to the group mean, then using different transformations to eliminate the skewness would also eliminate mean differences. It is analogous to an analysis of covariance, or a step-down analysis where we progressively covary out variables. We would be asking whether there were any between-group mean differences after controlling for skewness differences. If we do not want to ask such a question we must use the same transformation. In doing so we are implicitly assuming that the variables have the same distribution.

There are several common transformations used for reducing skewness, including the logarithmic transform. For this data set, we found that it led to a little improvement but that all variables still had positive skew. *Ad hoc* investigations led us instead to use $\ln(x-40)$, x being the raw observation. Table H.6 shows the results of the analysis after transforming the variables in this way.

Comparison of Tables H.5 and H.6 shows little difference between the results. The first (working upwards) significant effect is the condition by epoch interaction in both cases. However, closer examination of Table H.6 shows two other effects which almost achieve significance. It might be worthwhile

Psychophysiological responses

Table H.6 As Table H.5 but using observations transformed by $\ln(x-40)$. The hypothesis and error sums of products matrices have not been shown; the number of degrees of freedom associated with the error matrix is 56.

Source	d.f. for F	F	P
1 Grand mean (1 variable):			
Pref.	1/56	.914	.343
Cond.	1/56	.155	.695
Pref. cond.	1/56	3.728	.059

Source	d.f. for F	Approx. F	P
2 Epoch (2 variables):			
P E	2/55	1.365	.264
C E	2/55	3.460	.038
P C E	2/55	1.718	.189
3 Trial (3 variables):			
P T	3/54	.721	.544
C T	3/54	.525	.667
P C T	3/54	.825	.486
4 Epoch by trial (6 variables):			
P E T	6/51	1.746	.130
C E T	6/51	.699	.652
P C E T	6/51	2.148	.064

P: Preference: C: Condition. E: Epoch. T: Trial.

Table H.7 The univariate tests for the six variables used in testing the preference by condition interaction in block 4 of Table H.6.

Variable

Trial	Epoch	F	P
linear	× linear	0.387	.536
quadr.	× linear	3.691	.060
cubic	× linear	1.318	.256
linear	× quadr.	0.424	.518
quadr.	× quadr.	1.308	.258
cubic	× quadr.	6.953	.011

exploring these, perhaps with a view to setting up hypotheses for future work. The interpretation of the preference by condition interaction (seen to be almost significant at the 5% level using variable block 1, the grand mean variable) is straightforward. A few words about the preference by condition by epoch by trial interaction might be useful, however, since had this been significant we would have had to attempt to interpret it.

206

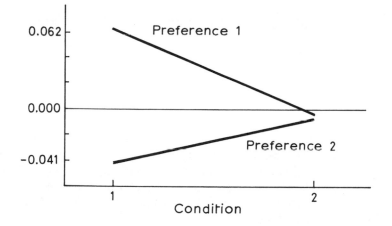

Figure H.6 The highest-order interaction variable shows a possible between-group inter-action effect.

The variables in block 4 summarize how the variation over epochs is modified by trials. The between-group interaction examines the way condition affects the change in this modification between preference levels. Had the effect been significant, the answer to the simple question 'Is there a difference between condition levels?' would have been 'Yes', in so far as the difference between preference levels varies between the condition levels.

Table H.7 shows the univariate F-tests for the six interaction variables in block 4. It seems that there may be a significant highest-order interaction component which is being diluted in the multivariate analysis by the other five variables. Figure H.6 illustrates this possible relationship. Of course, interpreting such a high-order effect is very difficult and we might be tempted to reject it on grounds of parsimony.

References

Aitkin, M. (1978) The analysis of unbalanced cross-classifications, *Journal of the Royal Statistical Society*, Series A, **141**, 195–223.

Bird, K. D. and Hadzi-Pavlovic, D. (1983) Simultaneous test procedures and the choice of a test statistic in MANOVA, *Psychological Bulletin*, **93**, 167–78.

Bock, R. D. (1975) *Multivariate Statistical Methods in Behavioural Research*, McGraw-Hill, New York.

Cochran, W. G. and Cox, G. M. (1957) *Experimental Designs*, Wiley, New York.

Dixon, W. J. *et al.* (ed.) (1981) *BMDP Statistical Software Manual*, University of California Press, Berkeley.

Finn, J. D. (1972) *MULTIVARIANCE: Univariate and Multivariate Analysis of Variance, Covariance and Regression*, National Educational Resources Inc., Ann Arbor, Michigan.

Finn, J. D. (1974) *A General Model for Multivariate Analysis*, Holt, Rinehart and Winston, New York.

Gabriel, K. R. (1968) Simultaneous test procedures in multivariate analysis of variance, *Biometrika*, **55**, 489–504.

Hand, D. J. (1981) Branch and bound in statistical data analysis, *The Statistician*, **30**, 1–13.

Harris, R. J. (1975) *A Primer of Multivariate Statistics*, Academic Press, New York.

Hays, W. L. (1963) *Statistics*, Holt, Rinehart and Winston, New York.

Hocking, R. R. and Speed, F. M. (1975) A full rank analysis of some linear model problems, *Journal of the American Statistical Association*, **70**, 706–12.

Kempthorne, O. (1952) *The Design and Analysis of Experiments*, Wiley, New York.

Keppel, G. (1973) *Design and Analysis: a Researcher's Handbook*, Prentice Hall, Englewood Cliffs, New Jersey.

McCabe, G. P. Jr. (1975) Computations for variable selection in discriminant analysis, *Technometrics*, **17**, 103–9.

McKay, R. J. (1977) Simultaneous procedures for variable selection in multiple discriminant analysis, *Biometrika*, **64**, 283–90.

McKay, R. J. and Campbell, N. A. (1982) Variable selection techniques in discriminant analysis 1: Description, *British Journal of Mathematical and Statistical Psychology*, **35**, 1–29.

References

Mardia, K. V., Kent, J. T., and Bibby, J. M. (1979) *Multivariate Analysis*, Academic Press, London.

Miller, R. G. Jr. (1981) *Simultaneous Statistical Inference* (2nd edn), Springer-Verlag, New York.

Morrison, D. F. (1967) *Multivariate Statistical Methods*, McGraw-Hill, New York.

Olson, C. L. (1974) Comparative robustness of six tests in multivariate analysis of variance, *Journal of the American Statistical Association*, **69**, 894–908.

Olson, C. L. (1976) On choosing a test statistic in multivariate analysis of variance, *Psychological Bulletin*, **83**, 579–85.

Olson, C. L. (1979) Practical considerations in choosing a MANOVA test statistic: a rejoinder to Stevens, *Psychological Bulletin*, **86**, 1350–2.

Rouanet, H. and Lépine, D. (1977) *Introduction à l'analyse des comparaisons pour le traitement des données expérimentales, Informatique et Science Humaine*, No. 33–4, Sorbonne, Paris.

Roy, J. (1958) Step-down procedure in multivariate analysis, *Annals of Mathematical Statistics*, **29**, 1177–87.

Roy, J. and Bargmann, R. E. (1958) Tests of multiple independence and the associated confidence bounds, *Annals of Mathematical Statistics*, **29**, 491–503.

SAS Institute Inc. (1982) *SAS User's Guide: Statistics*, SAS Institute Inc, Cary, North Carolina.

Scheffé, H. (1959) *The Analysis of Variance*, Wiley, New York.

SPSS Inc. (1983) *SPSS-X User's Guide*, McGraw-Hill, New York.

Srivastava, M. S. and Carter, E. M. (1983) *An Introduction to Applied Multivariate Statistics*, North-Holland, New York.

Stevens, J. (1979) Comment on Olson: choosing a test statistic in multivariate analysis of variance, *Psychological Bulletin*, **86**, 355–60.

Tabachnik, B. G. and Fidell, L. S. (1983) *Using Multivariate Statistics*, Harper and Row, New York.

Urquhart, N. S. and Weeks, D. L. (1978) Linear models in messy data: some problems and alternatives, *Biometrics*, **34**, 696–705.

Winer, B. J. (1971) *Statistical Principles in Experimental Design*, McGraw-Hill, Kogakusha, Tokyo.

Computer listings —————— **Appendix**

Appendix

INTRODUCTION

In this appendix we present some examples of computer programs and output listings. Our aim is to illustrate how the ideas in Part 1 of the book are defined in some of the standard computer program packages and how they appear in the output.

These examples are not intended to show how to use the computer packages themselves – clearly this much larger task is for the relevant manuals – but only to show how the control card sequences relate to the terminology in this book. Space allows no more than the spirit of a package's approach to be represented, but we hope to have caught in these examples the essence of those ideas.

The packages used for illustration are the SAS system, BMDP statistical software, and the SPSS-X batch system. These are each used to generate the tables of results presented in Study A in Part 2 of the book. Again, there is no intention to compare these packages as competing systems, nor to suggest that they are superior to others we have not used. We make no attempt to show these packages in their best light with regard to the ease of setting up their control card sequences – in every case the sequences we show can be simplified by reverting to default options built into the package – but rather we have a more basic intention. Our aim here is simply to make explicit the links between the ideas in Part 1 and the definitions used in the control card sequences. Needless to say, the output listings have been heavily edited to leave only the lines relevant to the points being made. They should not be taken as representing the fullness or brevity of the originals. At various places these listings have been marked at the start of a line with footnote indicators enclosed in square parentheses thus: [1]; otherwise they are direct copies of the computer listings. We should also remark that computer software evolves. By the time you read this book more-advanced versions of the programs used here may be available.

Output from SPSS-X procedures is printed with permission of SPSS Inc., Chicago, Illinois, 60611.

Output from SAS procedures is printed with permission of SAS Institute Inc., Cary, NC 27511-8000, Copyright © 1982.

Output from BMDP procedures is reproduced with permission of BMDP Statistical Software, Los Angeles, California, 90025.

Listing 1 SAS program for the analysis A.2 presented in Study A: a three-group between-subjects design (comprising one factor EXPER) with three variables in the within-subjects design (unstructured) (see Table A.5).

```
              DATA RPIDATA;
     [1]        INFILE DAT; INPUT UM ES LV EXPER ;

     [2]      PROC GLM DATA=RPIDATA;
     [3]        CLASS EXPER;
     [4]        MODEL UM ES LV = EXPER;
     [5]        CONTRAST 'HELMERT 1'  EXPER  1  -.5  -.5;
                CONTRAST 'HELMERT 2'  EXPER  0   1   -1;
     [6]        MANOVA  H=EXPER / PRINTH PRINTE;

              ENDSAS;
```

[1] Names, as used in the text, for the three variables and the group indicator EXPER (coded 1, 2 and 3).
[2] Call the GLM procedure, using the above variables and indicator.
[3] Declare the between-groups classification factor.
[4] Defining the simplest model – three unstructured response variables analysed by a single between-groups factor.
[5] Setting the contrasts used for the complete comparison of the between-groups factor EXPER. These are the formal coefficients of the Helmert contrasts discussed in the text (see Table A.6).
[6] The MANOVA command, with explicit print options.

Appendix

Listing 1a Output from the program in Listing 1: individual univariate analyses for each of the three variables.

GENERAL LINEAR MODELS PROCEDURE

CLASS LEVEL INFORMATION

CLASS	LEVELS	VALUES
EXPER	3	1 2 3

NUMBER OF OBSERVATIONS IN DATA SET = 251

[7]   GENERAL LINEAR MODELS PROCEDURE

DEPENDENT VARIABLE: UM

SOURCE	DF	SUM OF SQUARES	MEAN SQUARE	F VALUE
MODEL	2	841.76024790	420.88012395	3.41
ERROR	248	30636.62222222	123.53476703	PR > F
CORRECTED TOTAL	250	31478.38247012		0.0347

R-SQUARE	C.V.	ROOT MSE	UM MEAN
0.026741	56.1774	11.11461952	19.78486056

SOURCE	DF	TYPE I SS	F VALUE	PR > F
EXPER	2	841.76024790	3.41	0.0347

SOURCE	DF	TYPE III SS	F VALUE	PR > F
EXPER	2	841.76024790	3.41	0.0347

[8]

CONTRAST	DF	SS	F VALUE	PR > F
HELMERT 1	1	819.56809474	6.63	0.0106
HELMERT 2	1	21.34603175	0.17	0.6780

---

[7] Standard univariate analysis of variance of variable UM for the complete comparison of the factor EXPER.

[8] Note that the SS for the individual single-degree-of-freedom contrasts do not sum to the figure for the complete comparison. These are the unique SS for testing each contrast, without assuming anything about the value of the other.

214

Listing 1A (continued)

```
[9]                    GENERAL LINEAR MODELS PROCEDURE

DEPENDENT VARIABLE: ES

SOURCE                DF      SUM OF SQUARES        MEAN SQUARE      F VALUE
MODEL                  2        157.99353165       78.99676583         1.88
ERROR                248      10439.51244444       42.09480824       PR > F
CORRECTED TOTAL      250      10597.50597610                         0.1553

R-SQUARE           C.V.          ROOT MSE               ES MEAN
0.014909        57.8098         6.48805119           11.22310757

SOURCE                DF         TYPE I SS   F VALUE     PR > F
EXPER                  2        157.99353165    1.88     0.1553

SOURCE                DF       TYPE III SS   F VALUE     PR > F
EXPER                  2        157.99353165    1.88     0.1553

CONTRAST              DF                SS   F VALUE     PR > F
HELMERT 1              1        157.91232045    3.75     0.0539
HELMERT 2              1         13.82857143    0.33     0.5671

[10]                   GENERAL LINEAR MODELS PROCEDURE

DEPENDENT VARIABLE: LV

SOURCE                DF      SUM OF SQUARES        MEAN SQUARE      F VALUE
MODEL                  2        123.29888800       61.64944400         8.96
ERROR                248       1706.75688889        6.88208423       PR > F
CORRECTED TOTAL      250       1830.05577689                         0.0002

R-SQUARE           C.V.          ROOT MSE               LV MEAN
0.067374        45.4428         2.62337268            5.77290837

SOURCE                DF         TYPE I SS   F VALUE     PR > F
EXPER                  2        123.29888800    8.96     0.0002

SOURCE                DF       TYPE III SS   F VALUE     PR > F
EXPER                  2        123.29888800    8.96     0.0002

CONTRAST              DF                SS   F VALUE     PR > F
HELMERT 1              1        122.18985441   17.75     0.0001
HELMERT 2              1         20.31746032    2.95     0.0870
```

[9] Standard univariate analysis of variance of variable ES.
[10] Standard univariate analysis of variance of variable LV.

215

Appendix

Listing 1b Output from the program in Listing 1 (cont.). Multivariate analysis of the complete comparison.

```
                          GENERAL LINEAR MODELS PROCEDURE
[11]                           E = ERROR SS&CP MATRIX

    DF=248              UM                  ES                      LV
    UM            30636.62222222      10151.95555556          3878.24444444
    ES            10151.95555556      10439.51244444          2251.10977778
    LV             3878.24444444       2251.10977778          1706.75688889
```

PARTIAL CORRELATION COEFFICIENTS FROM THE ERROR SS&CP MATRIX / PROB > |R|

```
                 DF=247         UM       ES       LV
                 UM        1.000000 0.567661 0.536326
                           0.0000   0.0001   0.0001


                 ES        0.567661 1.000000 0.533299
                           0.0001   0.0000   0.0001


                 LV        0.536326 0.533299 1.000000
                           0.0001   0.0001   0.0000
```

```
                          GENERAL LINEAR MODELS PROCEDURE
[12]                      H = TYPE III SS&CP MATRIX FOR: EXPER

    DF=2                UM                  ES                      LV
    UM             841.76024790       361.09225321            311.49260735
    ES             361.09225321       157.99353165            138.60735370
    LV             311.49260735       138.60735370            123.29888800
```

[13]   CHARACTERISTIC ROOTS AND VECTORS OF: E INVERSE * H, WHERE
      H = TYPE III SS&CP MATRIX FOR: EXPER   E = ERROR SS&CP MATRIX

```
   CHARACTERISTIC   PERCENT     CHARACTERISTIC VECTOR   V'EV=1
      ROOT
                                     UM           ES            LV
     0.07387721      96.72      0.00087323   -0.00164303    0.02411623
     0.00250521       3.28     -0.00687964    0.00060679    0.01676509
     0.00000000       0.00     -0.00248449    0.01246841   -0.00773984
```

---

[11] Within-groups SP matrix, to be used in tests as the error SP matrix (see Table A.7).

[12] Between-groups hypothesis SP matrix for the complete comparison, with two degrees of freedom (see Table A.7). Note the diagonal elements are the univariate SS from Listing 1a.

[13] Eigenvalues of $(\mathbf{H \cdot E}^{-1})$ are also known as characteristic roots. Note that only two eigenvalues can be non-zero.

Listing 1b (continued)

```
[14]                    GENERAL LINEAR MODELS PROCEDURE

       MANOVA TEST CRITERIA FOR THE HYPOTHESIS OF NO OVERALL EXPER EFFECT

                         H = TYPE III SS&CP MATRIX FOR: EXPER
                         E = ERROR SS&CP MATRIX
                         P = RANK OF (H+E)     =        3
                         Q = HYPOTHESIS DF     =        2
                         NE= DF OF E           =      248
                         S = MIN(P,Q)          =        2
                         M = .5(ABS(P-Q)-1)    =      0.0
                         N = .5(NE-P)          =    122.5
[15]      --------------------------------------------------------
       WILKS' CRITERION       L = DET(E)/DET(H+E) =      0.92887813

               EXACT F = (1-SQRT(L))/SQRT(L)*(NE+Q-P-1)/P
                            WITH 2P AND 2(NE+Q-P-1) DF

                 F(6,492) =     3.08     PROB > F = 0.0057
       --------------------------------------------------------
       PILLAI'S TRACE         V = TR(H*INV(H+E)) =      0.07129378

           F APPROXIMATION = (2N+S)/(2M+S+1) * V/(S-V)
                            WITH S(2M+S+1) AND S(2N+S) DF

                 F(6,494) =     3.04     PROB > F = 0.0062
       --------------------------------------------------------
       HOTELLING-LAWLEY TRACE = TR(E**-1*H) =        0.07638241

           F APPROXIMATION = (2S*N-S+2)*TR(E**-1*H)/(S*S*(2M+S+1))
                            WITH S(2M+S+1) AND 2S*N-S+2 DF

                 F(6,490) =     3.12     PROB > F = 0.0052
[16]      --------------------------------------------------------
       ROY'S MAXIMUM ROOT CRITERION =                0.07387721

           FIRST CANONICAL VARIABLE YIELDS AN F UPPER BOUND

                 F(3,247) =     6.08     (UPPER BOUND)
```

[14] Multivariate test of the null hypothesis (that the complete comparison is zero).
[15] Four alternative statistics for testing the null hypothesis, as discussed in the text.
[16] Note that Roy's criterion is printed (by this version of SAS) as the largest eigenvalue of the matrix $(\mathbf{H}\cdot\mathbf{E}^{-1})$.

Appendix

Listing 1c Output from the program in Listing 1 (cont.). The separate multivariate testing of each contrast in the complete comparison.

```
                         GENERAL LINEAR MODELS PROCEDURE
[17]             H = CONTRAST SS&CP MATRIX FOR: HELMERT 1

     DF=1               UM                ES                LV
     UM           819.56809474     359.74977360      316.45363985
     ES           359.74977360     157.91232045      138.90746360
     LV           316.45363985     138.90746360      122.18985441

[18]      CHARACTERISTIC ROOTS AND VECTORS OF: E INVERSE * H, WHERE
     H = CONTRAST SS&CP MATRIX FOR: HELMERT 1     E = ERROR SS&CP MATRIX

     CHARACTERISTIC   PERCENT      CHARACTERISTIC  VECTOR     V'EV=1
         ROOT
                                      UM              ES              LV
       0.07342012    100.00      0.00097484     -0.00165182      0.02386577
       0.00000000      0.00      0.00523281      0.00460571     -0.01878806
       0.00000000      0.00     -0.00509235      0.01160120      0.00000000

                         GENERAL LINEAR MODELS PROCEDURE

     MANOVA TEST CRITERIA FOR THE HYPOTHESIS OF NO OVERALL HELMERT 1 EFFECT

                       H = CONTRAST SS&CP MATRIX FOR: HELMERT 1
                       E = ERROR SS&CP MATRIX
                       P = RANK OF (H+E)    =       3
                       Q = HYPOTHESIS DF    =       1
                       NE= DF OF E          =     248
                       S = MIN(P,Q)         =       1
                       M = .5(ABS(P-Q)-1)   =     0.5
                       N = .5(NE-P)         =   122.5
     ------------------------------------------------------------------
     WILKS' CRITERION      L = DET(E)/DET(H+E) =      0.93160169

             EXACT F = (1-L)/L*(NE+Q-P)/P
                            WITH P AND NE+Q-P DF

             F(3,246) =      6.02     PROB > F = 0.0006
     ------------------------------------------------------------------
     PILLAI'S TRACE        V = TR(H*INV(H+E)) =      0.06839831

             F APPROXIMATION = (2N+S)/(2M+S+1) * V/(S-V)
                            WITH S(2M+S+1) AND S(2N+S) DF

             F(3,246) =      6.02     PROB > F = 0.0006
     ------------------------------------------------------------------
     HOTELLING-LAWLEY TRACE = TR(E**-1*H) =      0.07342012

             F APPROXIMATION = (2S*N-S+2)*TR(E**-1*H)/(S*S*(2M+S+1))
                            WITH S(2M+S+1) AND 2S*N-S+2 DF

             F(3,246) =      6.02     PROB > F = 0.0006
     ------------------------------------------------------------------
     ROY'S MAXIMUM ROOT CRITERION =                 0.07342012

             FIRST CANONICAL VARIABLE YIELDS AN F UPPER BOUND

             F(3,246) =      6.02     PROB > F = 0.0006
```

Listing 1c (continued)

```
                          GENERAL LINEAR MODELS PROCEDURE
[19]                  H = CONTRAST SS&CP MATRIX FOR: HELMERT 2

 DF=1                    UM                   ES                   LV
 UM               21.34603175          17.18095238          20.82539683
 ES               17.18095238          13.82857143          16.76190476
 LV               20.82539683          16.76190476          20.31746032

           CHARACTERISTIC ROOTS AND VECTORS OF: E INVERSE * H, WHERE
       H = CONTRAST SS&CP MATRIX FOR: HELMERT 2     E = ERROR SS&CP MATRIX

      CHARACTERISTIC    PERCENT        CHARACTERISTIC VECTOR    V'EV=1
          ROOT
                                             UM              ES              LV
       0.01348925      100.00      -0.00192543     -0.00126789      0.02878630
       0.00000000        0.00      -0.00681595      0.00846831      0.00000000
       0.00000000        0.00       0.00202489      0.00923088     -0.00969099

                          GENERAL LINEAR MODELS PROCEDURE

   MANOVA TEST CRITERIA FOR THE HYPOTHESIS OF NO OVERALL HELMERT 2 EFFECT

                      H = CONTRAST SS&CP MATRIX FOR: HELMERT 2
                      E = ERROR SS&CP MATRIX
                      P = RANK OF (H+E)     =        3
                      Q = HYPOTHESIS DF     =        1
                      NE= DF OF E           =      248
                      S = MIN(P,Q)          =        1
                      M = .5(ABS(P-Q)-1)    =      0.5
                      N = .5(NE-P)          =    122.5
   ---------------------------------------------------------------
   WILKS' CRITERION        L = DET(E)/DET(H+E) =      0.98669029

                EXACT F = (1-L)/L*(NE+Q-P)/P
                              WITH P AND NE+Q-P DF

                   F(3,246) =      1.11      PROB > F = 0.3472
```

(Footnotes [17] to [19] appear on page 220.)

Appendix

Listing 1c (continued)

```
PILLAI'S TRACE          V = TR(H*INV(H+E)) =        0.01330971

       F APPROXIMATION = (2N+S)/(2M+S+1) * V/(S-V)
                         WITH S(2M+S+1) AND S(2N+S) DF

             F(3,246) =      1.11     PROB > F = 0.3472
---------------------------------------------------------------
HOTELLING-LAWLEY TRACE = TR(E**-1*H) =              0.01348925

       F APPROXIMATION = (2S*N-S+2)*TR(E**-1*H)/(S*S*(2M+S+1))
                         WITH S(2M+S+1) AND 2S*N-S+2 DF

             F(3,246) =      1.11     PROB > F = 0.3472
---------------------------------------------------------------
ROY'S MAXIMUM ROOT CRITERION =                     0.01348925

       FIRST CANONICAL VARIABLE YIELDS AN F UPPER BOUND

             F(3,246) =      1.11     PROB > F = 0.3472
```

[17] Hypothesis SP matrix for the 1st contrast used in constructing the complete comparison (see Table A.7).

[18] The single non-zero eigenvalue for the multivariate test of the single-degree-of-freedom contrast.

[19] The hypothesis SP matrix for the 2nd Helmert contrast, and its multivariate test (see Table A.7). As with the univariate analyses, the SP matrices for contrasts 1 and 2 do not sum to the overall hypothesis matrix for the complete comparison, since they represent the unique SP attributable to each contrast.

Listing 2 BMDP program for the analysis A.2 presented in Study A: a three-group between-subjects design (comprising one factor EXPER) with three variables in the within-subjects design (unstructured) (see Table A.5).

```
          / PROBLEM
                TITLE IS 'STUDY A'.
          / INPUT
                    VARIABLES ARE 4.
                    FORMAT IS FREE.
                    UNIT=10.
  [1]     / VARIABLE
                    NAMES ARE UM, ES, LV, EXPER.
  [2]     / BETWEEN
                    FACTOR IS EXPER.
                    CODES ARE 1 TO 3.
                    NAMES ARE LONG, RECENT, NEW.

          / WEIGHTS   BETWEEN ARE EQUAL.
                      WITHIN ARE EQUAL.

  [3]     / PRINT    MARGINALS ARE ALL.

          / END

  [4]     DESIGN    TYPE=BETWEEN, CONTRAST.
                    CODE=READ.
  [5]               VALUES = 1,  -1,  -1.  NAME = HELMERT1./
          DESIGN    VALUES = 0,   1,  -1.  NAME = HELMERT2./
          /
  [6]     ANALYSIS  PROCEDURE IS FACTORIAL.
                    EVALUES. DISPERSION.
  [7]               DEPENDENT ARE UM, ES, LV.
          /
```

[1] Names, as used in the text, for the three variables and the group indicator EXPER (coded 1, 2 and 3).

[2] Declare the between-groups classification factor.

[3] Request for means of variables used in the analysis.

[4] Setting the contrasts used for the complete comparison of the between-groups factor EXPER, partitioning the two degrees of freedom for a separate analysis for each. Note these values only show signs $(+ / 0 / -)$ for the Helmert contrast pattern. The formal coefficients discussed in the text (see Table A.6) are computed from them.

[6] Defining the simplest model – analysis by the (only) between-groups factor, with explicit print options for SP matrices.

[7] Analysis of the three unstructured response variables.

Appendix

Listing 2a Output from the program in Listing 2. The basic design information.

```
VARIABLES TO BE USED
    1 UM              2 ES          3 LV            4 EXPER

INPUT FORMAT IS
FREE
MAXIMUM LENGTH DATA RECORD IS   80 CHARACTERS.
[8]
BETWEEN FACTORS: EXPER

           | VARIABLE:
FACTOR:    | 1 UM        | 2 ES       | 3 LV        |
-----------------------------------------------------------
VARIATES |      1.000 |      2.000 |      3.000 |
         |      UM    |      ES    |      LV    |
-----------------------------------------------------------


NUMBER OF CASES READ. . . . . . . . . . . . .          251
[9]
WITHIN   DESIGN:
CELL NO. |  WEIGHT   |  SIZE   | VARIATES |
-------------------------------------------------
    1    | 1.00000 |     1   |        1   |
         |         |         |  UM        |
-------------------------------------------------
    2    | 1.00000 |     1   |        2   |
         |         |         |  ES        |
-------------------------------------------------
    3    | 1.00000 |     1   |        3   |
         |         |         |  LV        |
-------------------------------------------------


BETWEEN DESIGN:
CELL NO. |  WEIGHT   |  SIZE   | EXPER     |
-------------------------------------------------
    1    | 1.00000 |    90   |        2   |
         |         |         |  RECENT    |
-------------------------------------------------
    2    | 1.00000 |   125   |        1   |
         |         |         |  LONG      |
-------------------------------------------------
    3    | 1.00000 |    36   |        3   |
         |         |         |  NEW       |
```

[8] Statement that the between-subjects design has a single factor EXPER.
[9] The following two descriptions of the design (within and then between) do not themselves imply the existence of a factor structure for either. There is no within-subjects factor in the analysis.

Listing 2b Output from the program in Listing 2 (cont.). The requested analysis definition.

```
ANALYSIS CONTROL LANGUAGE

DESIGN    TYPE=BETWEEN, CONTRAST.
          CODE=READ.
          VALUES = 1,  -1,  -1.  NAME = HELMERT1./
*** WARNING --- MAKE SURE ORDER OF VALUES AGREES WITH ORDER OF
                CELLS IN DESIGN CHART.

DESIGN    VALUES = 0,  1,  -1.  NAME = HELMERT2./
*** WARNING --- MAKE SURE ORDER OF VALUES AGREES WITH ORDER OF
                CELLS IN DESIGN CHART.
/
ANALYSIS  PROCEDURE IS FACTORIAL.
          EVALUES. DISPERSION.
          DEPENDENT ARE UM, ES, LV.
/
FACTORIAL PROCEDURE .....
THE FOLLOWING STATEMENTS HAVE BEEN GENERATED:

  DESIGN  TYPE IS BETWEEN, CONTRAST.  MODEL.
    CODE IS CONST.  NAME IS 'OVALL: GRAND MEAN'./

  DESIGN  FACTOR IS EXPER.
    CODE IS EFFECT.  NAME IS 'E: EXPER'./

  ANALYSIS  /

END OF PROCEDURE-GENERATED STATEMENTS.
USER-SUPPLIED ANALYSIS CONTROL STATEMENTS RESUME FOLLOWING ANALYSIS OUTPUT.

========================================================================
--- ANALYSIS SUMMARY ---

THE FOLLOWING EFFECTS ARE COMPONENTS OF THE SPECIFIED
LINEAR MODEL FOR THE BETWEEN DESIGN.  ESTIMATES AND TESTS
OF HYPOTHESES FOR THESE EFFECTS CONCERN PARAMETERS OF THAT MODEL.
[10]      OVALL: GRAND MEAN
          E: EXPER

THE FOLLOWING EFFECTS INVOLVE WEIGHTED COMBINATIONS OF CELL MEANS.
THEY ARE NOT COMPONENTS OF THE LINEAR MODEL FOR THE BETWEEN DESIGN.
[11]      HELMERT1
          HELMERT2
```

[10] The between-subjects factor implies two 'effects': one degree of freedom for the grand mean, and two degrees of freedom for the effect of EXPER.
[11] Any linear combination of the means estimated under the (here saturated) model may be tested. The two Helmert contrasts have been selected.

Appendix

Listing 2c Output from the program in Listing 2 (cont.). The multivariate and univariate testing of each separate contrast in the complete comparison.

EFFECT	VARIATE	STATISTIC	F	DF	P
HELMERT1					
[12]	-ALL----				
		S= 1,T= 3,DFH= 1,DFE= 248			
		HT EVALS= 0.06839831			
		HE EVALS= 0.73420120D-01			
		TSQ= 18.2082	6.02	3, 246	0.0006
[13]	UM				
		SS= 8.1956809E+2			
		MS= 8.1956809E+2	6.63	1, 248	0.0106
	ES				
		SS= 1.5791232E+2			
		MS= 1.5791232E+2	3.75	1, 248	0.0539
	LV				
		SS= 1.2218985E+2			
		MS= 1.2218985E+2	17.75	1, 248	0.0000
HELMERT2					
[14]	-ALL----				
		S= 1,T= 3,DFH= 1,DFE= 248			
		HT EVALS= 0.01330971			
		HE EVALS= 0.13489246D-01			
		TSQ= 3.34533	1.11	3, 246	0.3472
[15]	UM				
		SS= 2.1346032E+1			
		MS= 2.1346032E+1	0.17	1, 248	0.6780
	ES				
		SS= 1.3828571E+1			
		MS= 1.3828571E+1	0.33	1, 248	0.5671
	LV				
		SS= 2.0317460E+1			
		MS= 2.0317460E+1	2.95	1, 248	0.0870

[12] The single non-zero eigenvalue for the multivariate test of the single degree of freedom contrast, Helmert1. Note that eigenvalues of both $(\mathbf{H} \cdot \mathbf{T}^{-1})$ and $(\mathbf{H} \cdot \mathbf{E}^{-1})$ matrices are given (see text).
[13] The separate univariate F-tests for the single-degree-of-freedom contrast.
[14] The eigenvalue for the multivariate test of the contrast Helmert2.
[15] The separate univariate F-tests for the single-degree-of-freedom contrast.

Listing 2d Output from the program in Listing 2 (cont.). The multivariate and univariate testing of the complete comparison.

```
OVALL: GRAND MEAN
[16]     -ALL----
               S=     1,T=    3,DFH=    1,DFE=    248
               HT EVALS= 0.79418709
               HE EVALS=    3.8587817
               TSQ=       956.978        316.42      3,    246  0.0000
[17]    UM
               SS=      6.9724371E+4
               MS=      6.9724371E+4     564.41      1,    248  0.0000
        ES
               SS=      2.2674151E+4
               MS=      2.2674151E+4     538.64      1,    248  0.0000
        LV
               SS=      5.6697740E+3
               MS=      5.6697740E+3     823.85      1,    248  0.0000
E: EXPER
[18]     -ALL----
               S=     2,T=    3,DFH=    2,DFE=    248
               HT EVALS= 0.06879483, 0.00249895
               HE EVALS=  0.73877205D-01,  0.25052067D-02
               LRATIO=  0.928878          3.08     6, 492.00 0.0057
               TRACE=   0.763824D-01
               TZSQ=     18.8665
                  CHISQ =   17.61                    5.576  0.0054
               MXROOT= 0.687948D-01                         0.0026
[19]    UM
               SS=      8.4176025E+2
               MS=      4.2088012E+2      3.41      2,    248  0.0347
        ES
               SS-      1.5799353E+2
               MS=      7.8996766E+1      1.88      2,    248  0.1553
        LV
               SS=      1.2329889E+2
               MS=      6.1649444E+1      8.96      2,    248  0.0002
[20]
 ERROR
        UM
               SS=      3.063662222E+4
               MS=     1.235347670E+2
        ES
               SS=      1.043951244E+4
               MS=      4.209480824E+1
        LV
               SS=      1.706756889E+3
               MS=      6.882084229E+0
```

[16] Multivariate test for the single-degree-of-freedom linear combination representing the grand mean vector of the sample (null hypothesis that the mean vector is zero).

[17] Univariate test of the hypotheses that the individual variables have zero means.

[18] Eigenvalues of $(\mathbf{H \cdot E}^{-1})$ and also $(\mathbf{H \cdot T}^{-1})$ matrices. Note that only two eigenvalues can be non-zero. Multivariate test of the null hypothesis (that the complete comparison for factor EXPER is zero) gives four alternative statistics for testing the null hypothesis, as discussed in the text. Note that Roy's criterion is given as the largest eigenvalue of the matrix $(\mathbf{H \cdot T}^{-1})$.

[19] Standard univariate tests of the hypotheses that the complete comparison for EXPER is zero (i.e. that there are no group differences on each of the three individual variables).

[20] Error mean squares used in univariate F-statistics' denominators.

Appendix

Listing 2d (continued)

```
[21]
    HYPOTHESIS DISPERSION MATRIX:
    HELMERT1
              1                2                3
    1   819.56809
    2   359.74977      157.91232
    3   316.45364      138.90746      122.18985

    HYPOTHESIS DISPERSION MATRIX:
    HELMERT2
              1                2                3
    1   21.346032
    2   17.180952       13.828571
    3   20.825397       16.761905       20.317460
[22]
    HYPOTHESIS DISPERSION MATRIX:
    OVALL: GRAND MEAN
              1                2                3
    1   69724.371
    2   39761.047       22674.151
    3   19882.692       11338.312       5669.7740
[23]
    HYPOTHESIS DISPERSION MATRIX:
    E: EXPER
              1                2                3
    1   841.76025
    2   361.09225      157.99353
    3   311.49261      138.60735      123.29889
[24]
    ERROR DISPERSION MATRIX:
              1                2                3
    1   30636.622
    2   10151.956       10439.512
    3   3878.2444       2251.1098       1706.7569
```

[21] Hypothesis SP matrices (used in above tests) for each of the two contrasts used in constructing the complete comparison (see Table A.7). As with the univariate analyses, the SP matrices for contrasts 1 and 2 do not sum to the overall hypothesis matrix for the complete comparison, since they represent the unique SP attributable to each contrast.

[22] Between-groups hypothesis SP matrix for the overall mean with one degree of freedom.

[23] And for the complete comparison with two degrees of freedom (see Table A.7). Note the diagonal elements are the univariate SS.

[24] Within-groups SP matrix, used in above tests as the error SP matrix (see Table A.7).

Listing 3 BDMP program for the analysis A.3 presented in Study A: a three-group between-subjects design (comprising one factor EXPER) with a three-variable within-subjects design (comprising one factor TYPE).

```
        / PROBLEM
                    TITLE IS 'STUDY A'.
        / INPUT
                    VARIABLES ARE 4.
                    FORMAT IS FREE.
                    UNIT=10.
  [1]   / VARIABLE
                    NAMES ARE UM, ES, LV, EXPER.
  [2]   / WITHIN
                    FACTOR IS TYPE.
                    CODES ARE 1 TO 3.
                    NAMES ARE UM, ES, LV.
        / BETWEEN
                    FACTOR IS EXPER.
                    CODES ARE 1 TO 3.
                    NAMES ARE LONG, RECENT, NEW.

        / WEIGHTS   BETWEEN ARE EQUAL.
                    WITHIN ARE EQUAL.

        / PRINT     MARGINALS ARE ALL.

        / END

  [3]   DESIGN   TYPE=BETWEEN, CONTRAST.
                 CODE=READ.
                 VALUES = 1,  -1,  -1.  NAME = HELMERT1./
        DESIGN   VALUES = 0,   1,  -1.  NAME = HELMERT2./
        /
  [4]   ANALYSIS PROCEDURE IS FACTORIAL.
                 EVALUES. DISPERSION.
        /
```

[1] Names, as used in the text, for the three variables and the group indicator EXPER (coded 1, 2 and 3).

[2] Declare (i) the within-subjects factor to structure the variables, and (ii) declare the between-groups classification factor.

[3] Setting the contrasts used for the complete comparison of the between-groups factor EXPER, partitioning the two degrees of freedom for a separate analysis for each. Note these values only show signs (+ / 0 / −) for the Helmert contrast pattern. The formal coefficients discussed in the text (see Table A.6) are computed from them. No particular contrasts have been defined here for the within-subjects factor TYPE.

[4] Defining the analysis: the within-subjects factor analysed by the between-groups factor, with explicit print options for SP matrices.

Appendix

Listing 3a Output from the program in Listing 3: the basic design information.

```
VARIABLES TO BE USED
    1 UM            2 ES         3 LV          4 EXPER

INPUT FORMAT IS
FREE
MAXIMUM LENGTH DATA RECORD IS   80 CHARACTERS.
[5]
BETWEEN FACTORS: EXPER
[6]
WITHIN FACTORS:  TYPE

        | VARIABLE:
FACTOR: |  1 UM     |  2 ES      |  3 LV      |
-----------------------------------------------------
TYPE    |    1.000  |    2.000   |    3.000   |
        |    UM     |    ES      |    LV      |
-----------------------------------------------------

NUMBER OF CASES READ. . . . . . . . . . . . .      251

WITHIN  DESIGN:
CELL NO. | WEIGHT  |  SIZE  | VARIATES |
-----------------------------------------------
   1  | 1.00000 |    1   |      1   |
      |         |        | UM       |
-----------------------------------------------
   2  | 1.00000 |    1   |      2   |
      |         |        | ES       |
-----------------------------------------------
   3  | 1.00000 |    1   |      3   |
      |         |        | LV       |
-----------------------------------------------

BETWEEN DESIGN:
CELL NO. | WEIGHT  |  SIZE  | EXPER    |
-----------------------------------------------
   1  | 1.00000 |   90   |      2   |
      |         |        | RECENT   |
-----------------------------------------------
   2  | 1.00000 |  125   |      1   |
      |         |        | LONG     |
-----------------------------------------------
   3  | 1.00000 |   36   |      3   |
      |         |        | NEW      |
-----------------------------------------------
```

[5] Statement that the within-subjects design has a single factor TYPE.
[6] Statement that the between-subjects design has a single factor EXPER.

Listing 3b Output from the program in Listing 2 (cont.): the requested analysis definition.

```
ANALYSIS CONTROL LANGUAGE

DESIGN   TYPE=BETWEEN, CONTRAST.
         CODE=READ.
         VALUES = 1,  -1,  -1.  NAME = HELMERT1./
*** WARNING --- MAKE SURE ORDER OF VALUES AGREES WITH ORDER OF
                CELLS IN DESIGN CHART.

DESIGN   VALUES = 0,   1,  -1.  NAME = HELMERT2./
*** WARNING --- MAKE SURE ORDER OF VALUES AGREES WITH ORDER OF
                CELLS IN DESIGN CHART.
/
[7]
ANALYSIS  PROCEDURE IS FACTORIAL.
          EVALUES. DISPERSION.
/

FACTORIAL PROCEDURE .....

THE FOLLOWING STATEMENTS HAVE BEEN GENERATED:
[8]
 DESIGN  TYPE IS BETWEEN, CONTRAST.  MODEL.
   CODE IS CONST.  NAME IS 'OVALL: GRAND MEAN'./

 DESIGN  FACTOR IS EXPER.
   CODE IS EFFECT.  NAME IS 'E: EXPER'./
[9]
 DESIGN  TYPE IS WITHIN, CONTRAST.  MODEL.
   CODE IS CONST.  NAME IS 'OBS: WITHIN CASE MEAN'./

 DESIGN  FACTOR IS TYPE.
   CODE IS EFFECT.  NAME IS 'T: TYPE'./

ANALYSIS  /

END OF PROCEDURE-GENERATED STATEMENTS.
USER-SUPPLIED ANALYSIS CONTROL STATEMENTS RESUME FOLLOWING ANALYSIS OUTPUT.
```

[7] Full factorial analysis requested; both the within and between factors will be used.

[8] The between-subjects factor implies two 'effects': one degree of freedom for the GRAND MEAN (i.e. the constant in the model) and two degrees of freedom for the complete comparison of the effect of EXPER. (The two Helmert contrasts have been previously selected.)

[9] The within-subjects factor implies that three new variables will be constructed from the original three: one representing the overall level of response for each subject ('WITHIN CASE MEAN'), and a set of two representing the complete comparison between levels of the factor TYPE. The program will pick default contrasts to construct these latter two variables (which will not be the same as those used for Table A.9(ii)).

Appendix

Listing 3b (continued)

```
--- ANALYSIS SUMMARY ---
[10]
THE FOLLOWING EFFECTS ARE COMPONENTS OF THE SPECIFIED
LINEAR MODEL FOR THE BETWEEN DESIGN.  ESTIMATES AND TESTS
OF HYPOTHESES FOR THESE EFFECTS CONCERN PARAMETERS OF THAT MODEL.

          OVALL: GRAND MEAN
          E: EXPER

THE FOLLOWING EFFECTS INVOLVE WEIGHTED COMBINATIONS OF CELL MEANS.
THEY ARE NOT COMPONENTS OF THE LINEAR MODEL FOR THE BETWEEN DESIGN.

          HELMERT1
          HELMERT2

THE FOLLOWING EFFECTS ARE COMPONENTS OF THE SPECIFIED
LINEAR MODEL FOR THE WITHIN  DESIGN.  ESTIMATES AND TESTS
OF HYPOTHESES FOR THESE EFFECTS CONCERN PARAMETERS OF THAT MODEL.

          OBS: WITHIN CASE MEAN
          T: TYPE

EFFECTS CONCERNING PARAMETERS OF THE COMBINED BETWEEN AND
WITHIN MODELS ARE THE COMBINATIONS (INTERACTIONS) OF EFFECTS
IN BOTH MODELS.
```

[10] The analysis proceeds with each of the between-group comparisons being analysed on each, in turn, of the two sets of variables in the between-subjects factor.

Listing 3c Output from the program in Listing 2 (contd.). Tests on the variable representing the overall response level.

```
WITHIN EFFECT:
[11]               OBS: WITHIN CASE MEAN
EFFECT    VARIATE       STATISTIC          F        DF         P
------------------------------------------------------------------------
 HELMERT1
 [12]    DEP_VAR
                    SS=    9.0996401E+2
                    MS=    9.0996401E+2     8.99    1,   248  0.0030
 HELMERT2
         DEP_VAR
                    SS=    5.5009524E+1
                    MS=    5.5009524E+1     0.54    1,   248  0.4618
 OVALL: GRAND MEAN
 [13]    DEP_VAR
                    SS=    8.0010799E+4
                    MS=    8.0010799E+4   790.07    1,   248  0.0000
 E: EXPER
 [14]    DEP_VAR
                    SS=    9.1514570E+2
                    MS=    4.5757285E+2     4.52    2,   248  0.0118
 ERROR
 [15]    DEP_VAR
                    SS=    2.511517037E+4
                    MS=    1.012708483E+2
```

[11] Analysis on the overall response level of each subject; sometimes called Subjects (or Observations) effect and in terminology of Repeated Measures Anova gives the Between Subjects Analysis. All tests use the single derived response variable 'WITHIN CASE MEAN' and are standard F-tests (see Table A.9(i) which uses corresponding normalized variable VO: the SS are three times greater but the F-values identical).

[12] Tests of the two partitioned Helmert contrasts on the derived response variable.

[13] Test of the grand mean, or constant, on the derived response variable. (Note that Table A.9(i) does not give 'unique' SS for grand mean, nor test it.) Univariate test of the hypothesis that the average response level has zero mean.

[14] Standard univariate test of the hypothesis that the complete comparison for EXPER is zero (i.e. that there are no group differences on the overall level of response). This can be considered as the main effect of factor EXPER.

[15] Error mean squares used in univariate F-statistics' denominators.

Appendix

Listing 3d Output from the program in Listing 2 (cont.). Tests of the separate between-group contrasts on the variables for the TYPE factor.

```
WITHIN EFFECT:
[16]              T: TYPE
EFFECT    VARIATE          STATISTIC              F        DF         P
----------------------------------------------------------------------
(T) X (HELMERT1)
[17]    DEP_VAR
                    S=   1,T=   2,DFH=   1,DFE=   248
                    HT EVALS= 0.01343602
                    HE EVALS=  0.13619001D-01
                    TSQ=        3.37751         1.68     2,   247  0.1881
               WCP SS=        1.8970626E+2
               WCP MS=        9.4853131E+1      2.66     2,   496  0.0707
               GREENHOUSE-GEISSER ADJ. DF       2.66  1.50, 370.86 0.0866
               HUYNH-FELDT ADJUSTED DF          2.66  1.51, 375.61 0.0859
(T) X (HELMERT2)
        DEP_VAR
                    S=   1,T=   2,DFH=   1,DFE=   248
                    HT EVALS= 0.00010581
                    HE EVALS=  0.10581714D-03
                    TSQ=      0.262427D-01      0.01     2,   247  0.9870
               WCP SS=        0.4825397E+0
               WCP MS=        0.2412698E+0      0.01     2,   496  0.9932
               GREENHOUSE-GEISSER ADJ. DF       0.01  1.50, 370.86 0.9791
               HUYNH-FELDT ADJUSTED DF          0.01  1.51, 375.61 0.9800
```

[16] Analysis on the two variables representing the complete comparison for the TYPE factor. In terminology of Repeated Measures Anova it gives the Within Subjects Analysis. Here all tests of the between-group comparisons are multivariate ones.

[17] Multivariate tests of each of the partitioned single-degree-of-freedom Helmert contrasts (see Table A.9(ii) – i.e. of the null hypothesis that the contrast shows zero differences between levels of TYPE. Note that with one degree of freedom for the comparison, all four multivariate tests would yield equivalent results – only one is shown.

232

Listing 3e Output from the program in Listing 2 (cont.). Tests on the variables representing the TYPE factor.

```
    T
[18]    DEP_VAR
                    S=   1,T=   2,DFH=  1,DFE=   248
                    HT EVALS= 0.61602676
                    HE EVALS=   1.6043482
                    TSQ=        397.878        198.14    2,    247  0.0000
              WCP SS=      1.8057497E+4
              WCP MS=      9.0287483E+3        253.47    2,    496  0.0000
              GREENHOUSE-GEISSER ADJ. DF       253.47  1.50, 370.86  0.0000
              HUYNH-FELDT ADJUSTED DF          253.47  1.51, 375.61  0.0000
   (T) X (E: EXPER)
[19]    DEP_VAR
                    S=   2,T=   2,DFH=  2,DFE=   248
[20]                HT EVALS= 0.01465732, 0.00009490
                    HE EVALS=  0.14875356D-01,  0.94908038D-04
                    LRATIO=  0.985249           0.92    4, 494.00  0.4513
                    TRACE=   0.149703D-01
                    TZSQ=       3.71263
                        CHISQ =    3.44                  3.763    0.4494
                    MXROOT=  0.146573D-01                         0.3854
              WCP SS=      2.0790697E+2
              WCP MS=      5.1976742E+1         1.46    4,    496  0.2134
              GREENHOUSE-GEISSER ADJ. DF        1.46  2.99, 370.86  0.2254
              HUYNH-FELDT ADJUSTED DF           1.46  3.03, 375.61  0.2250
   ERROR
           DEP_VAR
              WCP SS=      1.766772119E+4
              WCP MS=      3.562040562E+1

              GGI EPSILON      0.74771
              H-F EPSILON      0.75729
```

[18] Multivariate test for the grand mean vector on variables representing TYPE comparison (a single degree of freedom for the between-groups linear combination giving the sample grand mean). The null hypothesis is that the average TYPE differences are zero; note this can be considered as the main effect of factor TYPE.

[19] Multivariate tests for the complete comparison of EXPER on the variables for the TYPE comparison (null hypothesis is that any TYPE differences do not themselves differ between the EXPER groups). Note that this is often considered as the interaction of the within factor (TYPE) with the between factor (EXPER).

[20] Eigenvalues of $(\mathbf{H \cdot E}^{-1})$ and also $(\mathbf{H \cdot T}^{-1})$ matrices. Note that only two eigenvalues can be non-zero. Multivariate test of the null hypothesis gives four alternative statistics for testing the null hypothesis, as discussed in the text. Note that Roy's criterion is given as the largest eigenvalue of the matrix $(\mathbf{H \cdot T}^{-1})$.

Appendix

Listing 3f Output from the program in Listing 2 (cont.). Between-groups hypothesis SP matrices.

```
[21]
    HYPOTHESIS DISPERSION MATRIX:
    (T) X (HELMERT1)
               1              2
  1  77.212667
  2  6.6446311      0.57181191


  HYPOTHESIS DISPERSION MATRIX:
    (T) X (HELMERT2)
               1              2
  1 0.31746032D-02
  2-0.22222222D-01 0.15555556
[22]
    HYPOTHESIS DISPERSION MATRIX:
    T
               1              2
  1  8907.1904
  2  3552.4545      1416.8254
[23]
    HYPOTHESIS DISPERSION MATRIX:
    (T) X (E: EXPER)
               1              2
  1  85.518480
  2  8.5727950      1.0194281
[24]
    ERROR DISPERSION MATRIX:
               1              2
  1  6146.7226
  2  1432.3396      1911.0124
```

[21] Between-groups hypothesis SP matrices for the TYPE variables (used in above tests) for each of the two contrasts used in constructing the complete comparison of the between-groups factor EXPER. See [24] for the SP matrix for the complete comparison.

[22] Between-groups hypothesis SP matrix for the TYPE variables used in above test of the overall mean (with one degree of freedom) . . .

[23] . . . and in the test of the complete comparison for EXPER (with two degrees of freedom). As with the univariate analyses, the SP matrices for contrasts 1 and 2 do not sum to the overall hypothesis matrix for the complete comparison, since they represent the unique SP attributable to each contrast. (See [9] and note in Table A.9(ii) that the pair of variables V1, V2 used there have a different SP matrix, although the tests above are invariant to this re-expression of the variables.)

[24] Within-groups SP matrix, used in above tests as the error SP matrix.

Listing 4 SPSSX program for the analysis A.2 presented in Study A: a three-group between-subjects design (comprising one factor EXPER) with three variables in the within-subjects design (unstructured) (see Table A.5).

```
[1]    MANOVA      UM  ES  LV  BY EXPER(1,3)
[2]                /CONTRAST(EXPER)=HELMERT
[3]                /PARTITION(EXPER)
[4]                /PRINT=CELLINFO(MEANS)
                           SIGNIF(HYPOTH)
                           ERROR(SSCP)
                           DISCRIM( STAN ESTIM )
[5]                /METHOD=SSTYPE(SEQUENTIAL)
[6]                /ANALYSIS= UM  ES  LV
[7]                /DESIGN=EXPER
                   /DESIGN=EXPER(1) EXPER(2)
                   /DESIGN=EXPER(2) EXPER(1)
                   /DESIGN=EXPER(1)
       FINISH
```

[1] Call the MANOVA procedure using the variables and group indicator names, as used in the text, for the three variables and the classification factor EXPER (coded 1, 2 and 3).

[2] Setting the contrasts used for the complete comparison of the between-subjects factor EXPER. The formal coefficients of the Helmert contrasts discussed in the text (see Table A.6) are calculated automatically.

[3] Partition the between-groups comparison into single-degree-of-freedom (Helmert) contrasts.

[4] Take explicit print options.

[5] Request calculation of attributable SS an SP matrices sequentially, not the unique contribution of each contrast.

[6] Defining the simplest model – three unstructured response variables.

[7] Between-groups comparisons requested in four different ways:

    (i) by a single complete between-groups comparison;

    (ii) by sequentially testing two incomplete comparisons (Helmert1 and Helmert2 contrasts), constraining each, after testing it, to zero for all subsequent tests;

    (iii) by sequentially testing the incomplete comparisons in the reverse order;

    (iv) by a single incomplete comparison, Helmert1.

Appendix

Listing 4a Output from the program in Listing 4. The basic cell information for each variable.

```
* * * * * * * * * A N A L Y S I S   O F   V A R I A N C E * * * * * * * * * *
        251 cases accepted.
          0 cases rejected because of out-of-range factor values.
          5 cases rejected because of missing data.
          3 non-empty cells.
          4 designs will be processed.
- - - - - - - - - - - - - - - - - - - - - - - - - - - - - - - - - - - - - - -
[8]
Cell Means and Standard Deviations

Variable .. UM

        Factor            Code              Mean    Std. Dev.          N

    EXPER                   1              21.600    10.306          125
    EXPER                   2              18.244    11.477           90
    EXPER                   3              17.333    12.811           36

    For entire sample                      19.785    11.221          251


- - - - - - - - - - - - - - - - - - - - - - - - - - - - - - - - - - - - - -


Variable .. ES

        Factor            Code              Mean    Std. Dev.          N

    EXPER                   1              11.984     6.341          125
    EXPER                   2              10.678     6.563           90
    EXPER                   3               9.944     6.803           36

    For entire sample                      11.223     6.511          251

- - - - - - - - - - - - - - - - - - - - - - - - - - - - - - - - - - - - - -

Variable .. LV

        Factor            Code              Mean    Std. Dev.          N

    EXPER                   1               6.416     2.447          125
    EXPER                   2               5.389     2.680           90
    EXPER                   3               4.500     3.047           36

    For entire sample                       5.773     2.706          251
```

[8] Mean vectors for each of the three groups (but tabulated by variable) (see Table A.5).

Listing 4.1b Output from the program in Listing 4 for Design 1: The requested analysis definition and the within groups SP matrix.

```
Correspondence between Effects and Columns of BETWEEN-Subjects Design 1

     Starting  Ending
      Column   Column    Effect Name
[9]     1        1       CONSTANT
        2        3       EXPER
- - - - - - - - - - - - - - - - - - - - - - - - - - - - - - - - - - - - - -
* * * * * * * * * A N A L Y S I S   O F   V A R I A N C E * * * * * * * * *

Order of Variables for Analysis
[10]
    Variates      Covariates      Not Used
      UM
      ES
      LV

    3 Dependent Variables
    0 Covariates
    0 Variables not used
- - - - - - - - - - - - - - - - - - - - - - - - - - - - - - - - - - - - - -
WITHIN CELLS Sum-of-Squares and Cross-Products
[11]
                    UM            ES            LV
UM             30636.622
ES             10151.956     10439.512
LV              3878.244      2251.110     1706.757
```

---

[9] The between-subjects factor implies two 'effects': one degree of freedom for the grand mean, and two degrees of freedom for the effect of EXPER.

[10] Three raw response variables to be used in the analysis.

[11] The within-groups SP matrix for the three variables, to be used as the error SP matrix (see Table A.7).

Appendix

Listing 4.1c Output from the program in Listing 4 for Design 1 (cont.). The multivariate testing of the complete comparison for EXPER.

```
* * * * * * * * * A N A L Y S I S   O F   V A R I A N C E * * * * * * * * *
[12]
EFFECT .. EXPER
[13]
Adjusted Hypothesis Sum-of-Squares and Cross-Products
                      UM            ES            LV
    UM           841.760
    ES           361.092       157.994
    LV           311.493       138.607       123.299
- - - - - - - - - - - - - - - - - - - - - - - - - - - - - - - - - - - - -
Multivariate Tests of Significance (S = 2, M = 0, N = 122 )
[14]
Test Name           Value    Approx. F  Hypoth. DF    Error DF   Sig. of F
Pillais             .07129   3.04342        6.00        494.00       .006
Hotellings          .07638   3.11895        6.00        490.00       .005
Wilks               .92888   3.08137        6.00        492.00       .006

Roys largest root criterion =                .06879
- - - - - - - - - - - - - - - - - - - - - - - - - - - - - - - - - - - - -
Eigenvalues and Canonical Correlations
[15]
Root No.      Eigenvalue        Pct.    Cum. Pct. Canon. Cor.
    1             .074         96.720      96.720      .262
    2             .003          3.280     100.000      .050
- - - - - - - - - - - - - - - - - - - - - - - - - - - - - - - - - - - - -
Dimension Reduction Analysis

Roots       Wilks Lambda          F  Hypoth. DF    Error DF   Sig. of F
1 TO 2          .92888      3.08137        6.00      492.00        .006
2 TO 2          .99750       .30939        2.00      247.00        .734
- - - - - - - - - - - - - - - - - - - - - - - - - - - - - - - - - - - - -
Standardized discriminant function coefficients
[16]
               Function No.
Variable            1
UM                 .153
ES                -.168
LV                 .996
- - - - - - - - - - - - - - - - - - - - - - - - - - - - - - - - - - - - -
Estimates of effects for canonical variables

              Canonical Variable
   Parameter         1
        2          .568
        3          .331
```

Footnotes [12] to [16] appear on p. 241.

Listing 4.1d Output from the program in Listing 4 for Design 1 (cont.). The multivariate testing of the overall grand mean vector.

```
* * * * * * * * * * A N A L Y S I S   O F   V A R I A N C E * * * * * * * * *
[17]
EFFECT .. CONSTANT

Adjusted Hypothesis Sum-of-Squares and Cross-Products
                    UM           ES           LV
UM            98251.618
ES            55733.952    31615.494
LV            28668.263    16262.283     8364.944
- - - - - - - - - - - - - - - - - - - - - - - - - - - - - - - - - - - -
Multivariate Tests of Significance (S = 1, M = 1/2, N = 122 )
[18]
Test Name         Value    Approx. F  Hypoth. DF    Error DF   Sig. of F
Pillais           .84809   457.79381        3.00      246.00        .000
Hotellings       5.58285   457.79381        3.00      246.00        .000
Wilks             .15191   457.79381        3.00      246.00        .000

Roys largest root criterion =            .84809
- - - - - - - - - - - - - - - - - - - - - - - - - - - - - - - - - - - -
Eigenvalues and Canonical Correlations
[19]
Root No.   Eigenvalue      Pct.   Cum. Pct. Canon. Cor.
       1        5.583    100.000     100.000        .921
- - - - - - - - - - - - - - - - - - - - - - - - - - - - - - - - - - - -
Dimension Reduction Analysis

Roots    Wilks Lambda        F  Hypoth. DF    Error DF  Sig. of F
1 TO 1        .15191  457.79381        3.00      246.00       .000
- - - - - - - - - - - - - - - - - - - - - - - - - - - - - - - - - - - -
Standardized discriminant function coefficients

          Function No.
Variable             1
UM                -.270
ES                -.225
LV                -.672
- - - - - - - - - - - - - - - - - - - - - - - - - - - - - - - - - - - -
Estimates of effects for canonical variables

          Canonical Variable
Parameter            1
        1        -2.232
```

Footnotes [17] to [19] appear on p. 241.

Appendix

Listing 4.1e Output from the program in Listing 4 for Design 1 (cont.). The estimates of the overall mean (constant) and of the two Helmert contrast values in the complete comparison.

```
Estimates for UM
[20]
CONSTANT
  Parameter       Coeff.    Std. Err.    T-Value    Sig. of TLower 95% CL
        1   19.059259259      .80225    23.75733       .000    17.47917
  ParameterUpper 95% CL
        1       20.63935
[21]
EXPER
  Parameter       Coeff.    Std. Err.    T-Value    Sig. of TLower 95% CL
        2   3.8111111111     1.47963     2.57572       .011      .89686
        3    .9111111111     2.19183      .41568       .678    -3.40587
  ParameterUpper 95% CL
        2        6.72536
        3        5.22809
- - - - - - - - - - - - - - - - - - - - - - - - - - - - - - - - - -
Estimates for ES
[20]
CONSTANT
  Parameter       Coeff.    Std. Err.    T-Value    Sig. of TLower 95% CL
        1   10.868740741      .46830    23.20872       .000     9.94638
  ParameterUpper 95% CL
        1       11.79110
[21]
EXPER
  Parameter       Coeff.    Std. Err.    T-Value    Sig. of TLower 95% CL
        2   1.6728888889      .86372     1.93684       .054     -.02827
        3    .7333333333     1.27946      .57316       .567    -1.78666
  ParameterUpper 95% CL
        2        3.37405
        3        3.25333
- - - - - - - - - - - - - - - - - - - - - - - - - - - - - - - - - -
Estimates for LV
[20]
CONSTANT
  Parameter       Coeff.    Std. Err.    T-Value    Sig. of TLower 95% CL
        1    5.4349629630      .18935    28.70271       .000     5.06202
  ParameterUpper 95% CL
        1        5.80791
[21]
EXPER
  Parameter       Coeff.    Std. Err.    T-Value    Sig. of TLower 95% CL
        2   1.4715555556      .34924     4.21364       .000      .78371
        3    .8888888889      .51734     1.71820       .087     -.13004
  ParameterUpper 95% CL
        2        2.15940
        3        1.90782
```

[12] Sequentially computed SP matrix attributable to the complete comparison for EXPER (controlling, by default, for the grand mean across groups only). The sequential computation, in this case, gives the same as the unique computation.

[13] Between-groups hypothesis SP matrix for the complete comparison, with two degrees of freedom (see Table A.7). Note the diagonal elements are the univariate SS from Listing 1a.

[14] Multivariate tests of the null hypothesis (that the complete comparison is zero). Four alternative statistics for testing the null hypothesis, as discussed in the text. Note that Roy's criterion is given as the largest eigenvalue of the matrix $(\mathbf{H} \cdot \mathbf{T}^{-1})$.

[15] Eigenvalues of $(\mathbf{H} \cdot \mathbf{E}^{-1})$: note that only two eigenvalues can be non-zero. Canonical correlations for the comparison given (see text) and the Dimension Reduction Analysis, showing concentrated non-centrality (see text).

[16] Derived first canonical variable, and the values of the two Helmert contrasts calculated on this new variable.

[17] Between-groups hypothesis SP matrix for the overall mean with one degree of freedom, to be used in the test. Because of the sequential request, this is calculated constraining the EXPER comparison above to be zero (see Table A.7).

[18] Multivariate test for the single-degree-of-freedom linear combination representing the grand mean vector of the sample (null hypothesis that the mean vector is zero). Note that with a single degree of freedom all test criteria are equivalent.

[19] Eigenvalue of the $(\mathbf{H} \cdot \mathbf{E}^{-1})$ matrix. Note that only one eigenvalue can be non-zero.

[20] Estimate of the population grand mean (constant) vector (in conjunction with the unconstrained complete comparison), but tabulated by variable (see Table A.6).

[21] Estimate of the vector of population values for each of the two Helmert contrasts in the complete comparison, but tabulated by variable (see Table A.6).

Appendix

Listing 4.2a Output from the program in Listing 4, Design 2. The multivariate sequential testing of the separate contrasts in the comparison.

```
* * * * * * * * * A N A L Y S I S   O F   V A R I A N C E * * * * * * * * *
 Correspondence between Effects and Columns of BETWEEN-Subjects Design 2
[22]
   Starting  Ending
    Column   Column   Effect Name
      1        1      CONSTANT
      2        2      EXPER(1)
      3        3      EXPER(2)
- - - - - - - - - - - - - - - - - - - - - - - - - - - - - - - - - - - -
* * * * * * * * * A N A L Y S I S   O F   V A R I A N C E * * * * * * * * *
[23]
EFFECT .. EXPER(2)

Adjusted Hypothesis Sum-of-Squares and Cross-Products
                   UM           ES           LV
UM              21.346
ES              17.181       13.829
LV              20.825       16.762       20.317
- - - - - - - - - - - - - - - - - - - - - - - - - - - - - - - - - - - -
Multivariate Tests of Significance (S = 1, M = 1/2, N = 122 )

Test Name          Value   Approx. F  Hypoth. DF   Error DF  Sig. of F
Pillais           .01331   1.10612       3.00       246.00      .347
Hotellings        .01349   1.10612       3.00       246.00      .347
Wilks             .98669   1.10612       3.00       246.00      .347

Roys largest root criterion =            .01331
- - - - - - - - - - - - - - - - - - - - - - - - - - - - - - - - - - - -
Eigenvalues and Canonical Correlations
[24]
Root No.    Eigenvalue      Pct.    Cum. Pct. Canon. Cor.
    1          .013       100.000    100.000        .115
- - - - - - - - - - - - - - - - - - - - - - - - - - - - - - - - - - - -
Dimension Reduction Analysis

Roots     Wilks Lambda        F  Hypoth. DF   Error DF   Sig. of F
1 TO 1        .98669      1.10612     3.00      246.00        .347
- - - - - - - - - - - - - - - - - - - - - - - - - - - - - - - - - - - -
Standardized discriminant function coefficients

           Function No.
Variable        1
UM            -.337
ES            -.130
LV           1.189
- - - - - - - - - - - - - - - - - - - - - - - - - - - - - - - - - - - -
[25]
EFFECT .. EXPER(1)

Adjusted Hypothesis Sum-of-Squares and Cross-Products
                   UM           ES           LV
UM             820.414
ES             343.911      144.165
LV             290.667      121.845      102.981
- - - - - - - - - - - - - - - - - - - - - - - - - - - - - - - - - - - -
```

Listing 4.2a (continued)

```
* * * * * * * * * A N A L Y S I S   O F   V A R I A N C E * * * * * * * * *
Multivariate Tests of Significance (S = 1, M = 1/2, N = 122 )

Test Name          Value   Approx. F  Hypoth. DF    Error DF   Sig. of F
Pillais            .05917   5.15724       3.00       246.00       .002
Hotellings         .06289   5.15724       3.00       246.00       .002
Wilks              .94083   5.15724       3.00       246.00       .002

Roys largest root criterion =            .05917
- - - - - - - - - - - - - - - - - - - - - - - - - - - - - - - - - - - - -
Eigenvalues and Canonical Correlations
[26]
Root No.    Eigenvalue        Pct.   Cum. Pct. Canon. Cor.
       1          .063     100.000     100.000        .243
- - - - - - - - - - - - - - - - - - - - - - - - - - - - - - - - - - - - -
Dimension Reduction Analysis

Roots    Wilks Lambda        F  Hypoth. DF    Error DF   Sig. of F
1 TO 1         .94083   5.15724      3.00       246.00       .002
- - - - - - - - - - - - - - - - - - - - - - - - - - - - - - - - - - - - -
Standardized discriminant function coefficients

            Function No.
Variable           1
UM               .247
ES              -.172
LV               .939
- - - - - - - - - - - - - - - - - - - - - - - - - - - - - - - - - - - - -
Estimates of effects for canonical variables

            Canonical Variable
  Parameter          1
       2           .567
```

22] Sequential testing order for the two incomplete comparisons EXPER(1) and EXPER(2) (in this case the Helmert1 and Helmert2 contrasts, so each with one degree of freedom). EXPER(2) will be tested 'allowing for' EXPER(1) – i.e. in the complete comparison comprised of EXPER(1), EXPER(2). EXPER(1) will be tested after constraining the previously tested EXPER(2) to be zero (see text).

23] Hypothesis SP matrix and multivariate test of Helmert2 contrast, with one degree of freedom. The null hypothesis is that Helmert2 is zero, so Helmert1 (and the constant) will be adequate to account for the data. Note that, as it is part of a complete comparison, this is the same as the unique SP calculation. All four test criteria are equivalent because of the single degree of freedom for the incomplete comparison being tested (see Table A.7).

24] The eigenvalues of the $(\mathbf{H} \cdot \mathbf{E}^{-1})$ matrix – note that only one can be non-zero – and the associated canonical correlation and canonical variable.

25] Hypothesis SP matrix and multivariate test of Helmert1 contrast, with one degree of freedom, constraining the previously tested incomplete comparison (Helmert2) to be zero. Given that Helmert2 is zero, the null hypothesis is that Helmert1 is zero; so the 'constant' alone will be adequate to account for the data. Note that it is not part of a complete comparison. All four test criteria are equivalent because of the single degree of freedom for the incomplete comparison being tested.

26] The eigenvalues of the $(\mathbf{H} \cdot \mathbf{E}^{-1})$ matrix – note that only one can be non-zero – and the associated canonical correlation and canonical variable.

Appendix

Listing 4.2b Output from the program in Listing 4, Design 2 (cont.). The multivariate (sequential) testing of the overall grand mean vector.

```
* * * * * * * * * ANALYSIS   OF   VARIANCE * * * * * * * * *
[27]
EFFECT .. CONSTANT

Adjusted Hypothesis Sum-of-Squares and Cross-Products
                        UM            ES            LV
 UM              98251.618
 ES              55733.952     31615.494
 LV              28668.263     16262.283      8364.944
- - - - - - - - - - - - - - - - - - - - - - - - - - - - - - - -
Multivariate Tests of Significance (S = 1, M = 1/2, N = 122 )
. . . .
. . .
```

[27] This test is identical to that of Listing 4.1d.

Listing 4.3a Output from the program in Listing 4, Design 2. The multivariate sequential testing of the separate contrasts in the complete comparison.

```
* * * * * * * * * ANALYSIS   OF   VARIANCE * * * * * * * * *
 Correspondence between Effects and Columns of BETWEEN-Subjects Design 3
[28]
   Starting   Ending
    Column     Column    Effect Name
       1          1       CONSTANT
       2          2       EXPER(2)
       3          3       EXPER(1)
- - - - - - - - - - - - - - - - - - - - - - - - - - - - - - - -

. . . .
. . .
```

[28] This Design 3 request parallels exactly that of Design 2, except for the reversal of the two single degree-of-freedom contrasts in the sequential testing order. Thus the Helmert1 contrast estimated and tested as part of the complete comparison, yielding its 'unique' hypothesis S matrix (see Table A.7). The Helmert2 contrast is tested when the other is constrained to have zero value.

244

Listing 4.4a Output from the program in Listing 4, Design 4. The analysis of a single incomplete comparison for the factor EXPER.

```
* * * * * * * * * A N A L Y S I S   O F   V A R I A N C E * * * * * * * * *
Correspondence between Effects and Columns of BETWEEN-Subjects Design 4
[29]
   Starting  Ending
   Column    Column    Effect Name
      1         1       CONSTANT
      2         2       EXPER(1)
- - - - - - - - - - - - - - - - - - - - - - - - - - - - - - - - - - - - - -
* * * * * * * * * A N A L Y S I S   O F   V A R I A N C E * * * * * * * * *
[30]
EFFECT .. EXPER(1)

Adjusted Hypothesis Sum-of-Squares and Cross-Products
                  UM           ES           LV
UM             820.414
ES             343.911      144.165
LV             290.667      121.845      102.981
- - - - - - - - - - - - - - - - - - - - - - - - - - - - - - - - - - - - - -
Multivariate Tests of Significance (S = 1, M = 1/2, N = 122 )

Test Name          Value   Approx. F  Hypoth. DF  Error DF  Sig. of F
Pillais           .05917    5.15724      3.00      246.00      .002
Hotellings        .06289    5.15724      3.00      246.00      .002
Wilks             .94083    5.15724      3.00      246.00      .002

Roys largest root criterion =            .05917
- - - - - - - - - - - - - - - - - - - - - - - - - - - - - - - - - - - - - -
Eigenvalues and Canonical Correlations
[31]
Root No.    Eigenvalue      Pct.   Cum. Pct. Canon. Cor.
     1           .063     100.000   100.000      .243
- - - - - - - - - - - - - - - - - - - - - - - - - - - - - - - - - - - - - -
Dimension Reduction Analysis

Roots     Wilks Lambda       F  Hypoth. DF   Error DF  Sig. of F
1 TO 1        .94083    5.15724     3.00      246.00      .002
- - - - - - - - - - - - - - - - - - - - - - - - - - - - - - - - - - - - - -
Standardized discriminant function coefficients

           Function No.
Variable        1
UM             .247
ES            -.172
LV             .939
- - - - - - - - - - - - - - - - - - - - - - - - - - - - - - - - - - - - - -
Estimates of effects for canonical variables

           Canonical Variable
 Parameter          1
     2            .499
```

[29] Only the 'constant' and the incomplete comparison EXPER(1) are used instead of the complete comparison comprising EXPER(1), EXPER(2). EXPER(1) will be estimated and tested after constraining the omitted part of the complete comparison – EXPER(2) – to be zero (see text). Footnotes [30] and [31] appear on p. 246.

Listing 4.4b Output from the program in Listing 4, Design 4 (cont.). The estimation of a single incomplete comparison for the factor EXPER.

```
Estimates for UM
[32]
CONSTANT

    Parameter       Coeff.    Std. Err.    T-Value    Sig. of TLower 95% CL
          1  19.189417989      .73862     25.98014      .000      17.73465
    ParameterUpper 95% CL
          1       20.64418
[33]
EXPER(1)
    Parameter       Coeff.    Std. Err.    T-Value    Sig. of TLower 95% CL
          2   3.6158730159    1.40311      2.57704      .011       .85235
    ParameterUpper 95% CL
          2        6.37940
- - - - - - - - - - - - - - - - - - - - - - - - - - - - - - - - - - - - -
Estimates for ES
[32]
CONSTANT

    Parameter       Coeff.    Std. Err.    T-Value    Sig. of TLower 95% CL
          1  10.973502646      .43116     25.45103      .000      10.12430
    ParameterUpper 95% CL
          1       11.82271
[33]
EXPER(1)
    Parameter       Coeff.    Std. Err.    T-Value    Sig. of TLower 95% CL
          2   1.5157460317     .81905      1.85061      .065      -.09744
    ParameterUpper 95% CL
          2        3.12893
- - - - - - - - - - - - - - - - - - - - - - - - - - - - - - - - - - - - -
Estimates for LV
[32]
CONSTANT

    Parameter       Coeff.    Std. Err.    T-Value    Sig. of TLower 95% CL
          1   5.5619470899     .17434     31.90371      .000       5.21858
    ParameterUpper 95% CL
          1        5.90531
[33]
EXPER(1)
    Parameter       Coeff.    Std. Err.    T-Value    Sig. of TLower 95% CL
          2   1.2810793651     .33117      3.86829      .000       .62881
    ParameterUpper 95% CL
          2        1.93335
```

[30] Hypothesis SP matrix and multivariate test of Helmert1 contrast, with one degree of freedom, constraining the previously tested incomplete comparison (Helmert1) to be zero. The null hypothesis is that, given Helmert2 is zero, Helmert1 is zero; so the 'constant' alone will be adequate to account for the data. Note that it is an identical test to that in [25] above, because of the sequential calculations there.

[31] The eigenvalues of the $(\mathbf{H \cdot E}^{-1})$ matrix – note that only one can be non-zero – and the associated canonical correlation and canonical variable. Note that they are identical to figures given in [26], because of the sequential calculations there.

[32] Estimate of the population grand mean (constant) vector (in conjunction with only the incomplete comparison EXPER(1)), but tabulated by variable.

[33] Estimate of the vector of population values for the single contrast Helmert1 comprising this incomplete comparison, but tabulated by variable. Note these differ from the estimates associated with the complete comparison in [20], [21] (see Table A.8).

Listing 5 SPSSX program for the analysis A.2 presented in Study A: a three-group between-subjects design (comprising one factor EXPER) with three variables in the within-subjects design (comprising a single factor TYPE).

```
[1]    MANOVA      UM  ES  LV  BY  EXPER(1,3)
[2]                /WSFACTORS=TYPE(3)
[3]                /TRANSFORM( UM ES LV ) = SPECIAL
                            ( 1  1  1 , 1 -1  0 , 1  0 -1 )
[4]                /RENAME=VO V1 V2
[5]                /CONTRAST(EXPER)=HELMERT
[6]                /PARTITION(EXPER)
[7]                /PRINT=TRANSFORM
                        CELLINFO(MEANS)
                        SIGNIF(HYPOTH)
                        ERROR(SSCP)
                        DISCRIM( STAN ESTIM )
[8]                /METHOD=SSTYPE(SEQUENTIAL)
[9]                /ANALYSIS=(VO/V1 V2)
[10]               /DESIGN=EXPER
                   /DESIGN=EXPER(1) EXPER(2)
                   /DESIGN=EXPER(2) EXPER(1)
                   /DESIGN=EXPER(1)
       FINISH
```

[1] Call the MANOVA procedure using the variables and group indicator names, as used in the text, for the three variables and the classification factor EXPER (coded 1, 2 and 3).

[2] Declare the within-subjects factor TYPE for the three response variables.

[3] Set definitions to derive prespecified combinations of variables: the coefficients for the overall level of response; and the coefficients for two new variables to represent the complete comparison of the within-subjects factor TYPE (see text).

[4] Names for the new, transformed variables as used in the text.

[5] Setting the contrasts used for the complete comparison of the between-subjects factor EXPER. The formal coefficients of the Helmert contrasts discussed in the text (see Table A.6) are calculated automatically.

[6] Partition the between-groups comparison into single-degree-of-freedom (Helmert) contrasts.

[7] Take explicit print options.

[8] Request calculation of attributable SS an SP matrices sequentially, note the unique contribution of each contrast.

[9] Request analysis separately on the two sets of variables V0 and V1, V2.

[10] Between-groups comparisons requested in four different ways:
    (i) by a single complete between-groups comparison;
    (ii) by sequentially testing two incomplete comparisons (Helmert1 and Helmert2 contrasts), constraining each, after testing it, to zero for all subsequent tests;
    (iii) by sequentially testing the incomplete comparisons in the reverse order;
    (iv) by a single incomplete, Helmert1.

Appendix

Listing 5.1a Output from the program in Listing 5, Design 1. The within-subjects design and the univariate analysis using variable V0.

```
* * * * * * * * * A N A L Y S I S   O F   V A R I A N C E * * * * * * * * * *
Transformation Matrix (Transposed)
[11]
                  1          2          3
        1       1.000      1.000      1.000
        2       1.000     -1.000       .000
        3       1.000       .000     -1.000
- - - - - - - - - - - - - - - - - - - - - - - - - - - - - - - - - - - - -
Correspondence between Effects and Columns of BETWEEN-Subjects Design 1
[12]
   Starting  Ending
   Column    Column   Effect Name
      1         1      CONSTANT
      2         3      EXPER
- - - - - - - - - - - - - - - - - - - - - - - - - - - - - - - - - - - - -
* * * * * * * * * A N A L Y S I S   O F   V A R I A N C E * * * * * * * * * *

Order of Variables for Analysis
[13]
   Variates      Covariates      Not Used
   *V0                           V1
                                 V2

   1 Dependent Variable
   0 Covariates
   2 Variables not used
- - - - - - - - - - - - - - - - - - - - - - - - - - - - - - - - - - - - -
* * * * * * * * * A N A L Y S I S   O F   V A R I A N C E * * * * * * * * * *
Tests of significance for V0 using SEQUENTIAL sums of squares
[14]
Source of Variation          SS        DF       MS         F     Sig of F
WITHIN CELLS             75345.511    248    303.813
CONSTANT               339561.052          1339561.052  1117.666    .000
EXPER                     2745.437      2   1372.719     4.518     .012
- - - - - - - - - - - - - - - - - - - - - - - - - - - - - - - - - - - - -
Estimates for V0
[15]
CONSTANT
  Parameter       Coeff.    Std. Err.    T-Value   Sig. of TLower 95% CL
          1   35.362962963   1.25810    28.10814      .000    32.88503
  ParameterUpper 95% CL
          1       37.84089
[16]
EXPER
  Parameter       Coeff.    Std. Err.    T-Value   Sig. of TLower 95% CL
          2   6.9555555556   2.32040     2.99757      .003     2.38536
          3   2.5333333333   3.43729      .73702      .462    -4.23666
  ParameterUpper 95% CL
          2      11.52575
          3       9.30333
```

Footnotes [11] to [16] appear on p. 250.

Listing 5.1b Output from the program in Listing 5, Design 1 (cont.). The multivariate analysis using variables V1, V2.

```
* * * * * * * * * A N A L Y S I S   O F   V A R I A N C E * * * * * * * * *

Order of Variables for Analysis
[17]
     Variates      Covariates      Not Used
     *V1                           VO
     *V2

     2 Dependent Variables
     0 Covariates
     1 Variable not used
- - - - - - - - - - - - - - - - - - - - - - - - - - - - - - - - - - -
WITHIN CELLS Sum-of-Squares and Cross-Products
[18]
                    V1            V2
V1           20772.224
V2           18857.532    24586.890
- - - - - - - - - - - - - - - - - - - - - - - - - - - - - - - - - - -
* * * * * * * * * A N A L Y S I S   O F   V A R I A N C E * * * * * * * * *

EFFECT .. EXPER
[19]
Adjusted Hypothesis Sum-of-Squares and Cross-Products
                    V1            V2
V1             277.569
V2             307.783      342.074
- - - - - - - - - - - - - - - - - - - - - - - - - - - - - - - - - - -
Multivariate Tests of Significance (S = 2, M = -1/2, N = 122 1/2)
[20]
Test Name        Value    Approx. F  Hypoth. DF   Error DF   Sig. of F
Pillais         .01475     .92143        4.00      496.00       .451
Hotellings      .01497     .92067        4.00      492.00       .452
Wilks           .98525     .92107        4.00      494.00       .451

Roys largest root criterion =            .01466
- - - - - - - - - - - - - - - - - - - - - - - - - - - - - - - - - -
Eigenvalues and Canonical Correlations
[21]
Root No.    Eigenvalue      Pct.    Cum. Pct. Canon. Cor.
      1          .015      99.366     99.366       .121
      2          .000        .634    100.000       .010
- - - - - - - - - - - - - - - - - - - - - - - - - - - - - - - - - -
Dimension Reduction Analysis

Roots    Wilks Lambda        F  Hypoth. DF   Error DF   Sig. of F
1 TO 2        .98525     .92107      4.00      494.00       .451
2 TO 2        .99991     .02354      1.00      248.00       .878
```

Footnotes [17] to [21] appear on p. 251.

Appendix

Listing 5.1b (continued)

```
* * * * * * * * * A N A L Y S I S    O F    V A R I A N C E * * * * * * * * * *
EFFECT .. CONSTANT
[22]
Adjusted Hypothesis Sum-of-Squares and Cross-Products
                     V1              V2
V1              18399.207
V2              30111.685    49280.036
- - - - - - - - - - - - - - - - - - - - - - - - - - - - - - - - - - - - - -
Multivariate Tests of Significance (S = 1, M = 0, N = 122 1/2)
[23]
Test Name          Value    Approx. F  Hypoth. DF    Error DF   Sig. of F
Pillais            .68694   270.99281      2.00        247.00      .000
Hotellings        2.19427   270.99281      2.00        247.00      .000
Wilks              .31306   270.99281      2.00        247.00      .000

Roys largest root criterion =             .68694
- - - - - - - - - - - - - - - - - - - - - - - - - - - - - - - - - - - - - -
Eigenvalues and Canonical Correlations
[24]
Root No.     Eigenvalue      Pct.    Cum. Pct. Canon. Cor.
      1         2.194      100.000     100.000        .829
- - - - - - - - - - - - - - - - - - - - - - - - - - - - - - - - - - - - - -
Dimension Reduction Analysis

Roots     Wilks Lambda        F  Hypoth. DF    Error DF   Sig. of F
1 TO 1        .31306   270.99281      2.00        247.00      .000
```

[11] The transformation matrix to construct the derived variables from the originals, as in [3].

[12] The between-subjects factor implies two 'effects': one degree of freedom for the grand mean, and two degrees of freedom for the effect of EXPER.

[13] Analysis on the overall response level of each subject; sometimes called Subjects (or Observations) effect and in terminology of Repeated Measures Anova gives the Between Subjects Analysis.

[14] All tests use the single derived response variable V0 and are standard F-tests (see Table A.9(i)):
  (i) Univariate test of the grand mean, or constant, on the derived response variable (null hypothesis that the average response level has zero mean).
  (ii) Standard univariate test of the hypothesis that the complete comparison for EXPER is zero (i.e. that there are no group differences on the overall level of response). This can be considered as the main effect of factor EXPER.

[15] Estimates of the grand mean (constant) for the overall subject response level. This can be considered as the constant for the overall (between and within) factorial design.

[16] Estimates of the values of the two Helmert contrasts comprising the complete comparison for EXPER for the overall subject response level. These are the EXPER main effects.

[17] Analysis on the two variables representing the complete comparison for the TYPE factor, V1 and V2. In terminology of Repeated Measures Anova it gives the Within Subjects Analysis. Here all tests of the between-group comparisons are multivariate ones.

[18] Within-groups SP matrix, used in the tests as the error SP matrix.

[19] Between-groups hypothesis SP matrices using the TYPE variables in the test of the complete comparison for EXPER (Table A.9(ii)).

[20] Multivariate tests for the complete comparison of EXPER on the variables for the TYPE comparison (null hypothesis is that any TYPE differences do not themselves differ between the EXPER groups). Note that this is often considered as the interaction of the within factor (TYPE) with the between factor (EXPER).

Listing 5.1c Output from the program in Listing 5, Design 1 (cont.). The estimates for the between group comparison contrasts and the grand mean.

```
Estimates for V1
[25]
CONSTANT
  Parameter       Coeff.    Std. Err.    T-Value    Sig. of TLower 95% CL
        1  8.1905185185      .66059     12.39887       .000      6.88944
  ParameterUpper 95% CL
        1        9.49159
[26]
EXPER
  Parameter       Coeff.    Std. Err.    T-Value    Sig. of TLower 95% CL
        2  2.1382222222     1.21836      1.75500       .080      -.26143
        3   .1777777778     1.80480       .09850       .922     -3.37691
  ParameterUpper 95% CL
        2        4.53787
        3        3.73246
- - - - - - - - - - - - - - - - - - - - - - - - - - - - - - - - - - - - -
Estimates for V2
[25]
CONSTANT
  Parameter       Coeff.    Std. Err.    T-Value    Sig  of TLower 95% CL
        1  13.624296296      .71869     18.95721       .000     12.20879
  ParameterUpper 95% CL
        1       15.03980
[26]
EXPER
  Parameter       Coeff.    Std. Err.    T-Value    Sig. of TLower 95% CL

        2  2.3395555556     1.32552      1.76501       .079      -.27115
        3   .0222222222     1.96354       .01132       .991     -3.84511
  ParameterUpper 95% CL
        2        4.95026
        3        3.88955
```

[21] Eigenvalues of $(\mathbf{H \cdot E}^{-1})$ matrix and the associated canonical correlations and Dimension Reduction Analysis. Note that only two eigenvalues can be non-zero.

[22] Between-groups hypothesis SP matrix using the TYPE variables in the test of the overall mean, or 'constant' (with one degree of freedom).

[23] Multivariate test for the grand mean vector on variables representing TYPE comparison (a single degree of freedom for the between-groups linear combination giving the sample grand mean). The null hypothesis is that the average TYPE differences are zero; note this can be considered as the main effect of factor TYPE.

[24] Eigenvalues of $(\mathbf{H \cdot E}^{-1})$ matrix and the associated canonical correlations and Dimension Reduction Analysis. Note that only one eigenvalue can be non-zero, since there is only one between-groups degree of freedom.

[25] Estimates of the grand mean ('constant') vector for the variables representing the TYPE comparison, but tabulated by variable. These are the TYPE main effects.

[26] Estimates of the values of the two Helmert contrasts comprising the complete comparison for EXPER, tabulated by the variables representing the TYPE comparison. Note that these can be considered as interaction effects.

Index

Index

Index

Index

Index

Index

7859